Student's Solutions Manual

Prealgebra
A Transitional Approach

STUDENT'S SOLUTIONS MANUAL

PREALGEBRA
A TRANSITIONAL APPROACH

Jill Beer

Dwyn Peake

Kent State University

 **ADDISON-WESLEY**

An imprint of Addison Wesley Longman, Inc.

Reading, Massachusetts • Menlo Park, California • New York • Harlow, England
Don Mills, Ontario • Sydney • Mexico City • Madrid • Amsterdam

Reproduced by Addison-Wesley Educational Publishers Inc. from camera-ready copy supplied by the author.

Copyright © 1997 Addison-Wesley Educational Publishers Inc.

All rights reserved. No part of this publication may be reproduced, stored in a retrieval system, or transmitted, in any form or by any means, electronic, mechanical, photocopying, recording, or otherwise, without the prior written permission of the publisher. Printed in the United States of America.

ISBN   0-321-01274-7

3 4 5 6 7 8 9 10 VG 99

Reprinted with corrections, April 1999

CONTENTS

Solutions

STUDENT'S SOLUTIONS MANUAL

PREALGEBRA
A TRANSITIONAL APPROACH

CHAPTER 1: SETS

1.1 to 1.4

1. The subsets of {0,1,2} are: {0,1,2}, {0,1}, {0,2}, {1,2}, {0}, {1}, {2}, {}

2. A∪B = {0,1,2} ∪ {1,2,3,4} = {0,1,2,3,4}

3. A∩B∩C = {0,1,2} ∩ {1,2,3,4}∩ {4,5} = {} or ∅

4. A∪(B∩C) = {0,1,2} ∪ ({1,2,3,4} ∩ {4,5}
   = {0,1,2} ∪ {4}
   = {0,1,2,4}

5. The subsets of {1,2,3,4} are: {1,2,3,4}, {1,2,3}, {1,2,4}, {1,3,4}, {2,3,4}, {1,2}, {1,3}, {1,4}, {2,3}, {2,4}, {3,4}, {1}, {2} {3} {4} {}

6. A∩B = {0,1,2} ∩ {1,2,3,4} = {1,2}

7. A∪B∪C = {0,1,2} ∪ {1,2,3,4} ∪ {4,5}
   = {0,1,2,3,4,5}

8. A∩(B∪C) = {0,1,2} ∩ ({1,2,3,4} ∪ {4,5}
   = {0,1,2} ∩ {1,2,3,4,5}
   = {1,2}

9.

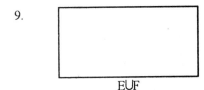

E∪F

10.

E∩F∩G

11.

12.

E∩F

E∪F∪G

13. The set of even numbers {2,4,6 . . .} is infinite.

14. The set {0,1,2,3,4,5,6,7,8,9} is finite.

15. The set of natural numbers is {1,2,3 . . .}
    The set of whole numbers is (0,1,2,3 . . .}
    {1,2,3 . . .} ∩ {0,1,2,3 . . .} = {1,2,3 . . .}

16. The set of natural numbers is {1,2,3 . . .}
    The set of whole numbers is {0,1,2,3 . . .}; {1,2,3 . . .} ∪ {0,1,2,3 . . .} = {0,1,2,3 . . .}

17. K⊆L true

18. L⊆K false L⊄K

19. M⊆K true

20. M ⊆S true

21. L⊆S true

22. S⊆M false S⊄M

23. M⊆S true

24. P⊆M false P⊄M

25. 2∈K true

26. ∅⊆P true

27. S=L true

REVIEW

1. See Section 1.1
2. See Section 1.1
3. See Section 1.2
4. See Section 1.1
5. See Section 1.2
6. See Section 1.1
7. See Section 1.3
8. See Section 1.3

1

10.	See Section 1.3
11.	See Section 1.4
12.	See Section 1.4
13.	See Section 1.4
14.	∈ is an element of
15.	∉ is not an element of
16.	∅ empty set
17.	∪ union
18.	∩ intersection
19.	⊆ is a subset of
20.	⊄ is not a subset of
21.	{} empty set
22.	= equal

PRACTICE TEST

1. The set of natural numbers is {1,2,3 . . .}

2. The set of whole numbers is {0,1,2,3 . . .}

3. The set of the letters of the alphabet is {a,b,c . . . x,y,z}

4. F∩G = {1,2,3}∩{3,4,5} = {3}

5. F∪H = {1,2,3}∪{1,4} = {1,2,3,4}

6. F∪(G∩H) = {1,2,3}∪({3,4,5}∩{1,4})

    = {1,2,3}∪{4}
    = {1,2,3,4}

7.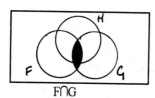

    F∩G

8.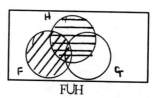

    F∪H

9.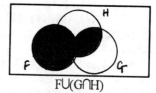

    F∪(G∩H)

10. ∅⊆F true

11. G⊄F true

12. H⊆H true

13. 3∈G true

14. 2∈H false 2∉H

15. 4∉F true

16. The subsets of {a,b,c,d} are: {a,b,c,d} {a,b,c} {a,b,d} {b,c,d} {a,c,d} {a,b} {a,c} {a,d} {b,c} {b,d} {c,d} {a} {b} {c} {d} {}

17. The set of letters of the alphabet is finite. The set of whole numbers is infinite.

18. Answers will vary.

2

CHAPTER 2: WHOLE NUMBERS

2.1 Place Value and Expanded Notation

1. $23 = 2 \times 10 + 3 \times 1$
2. $45 = 4 \times 10 + 5 \times 1$
3. $789 = 7 \times 100 + 8 \times 10 + 9 \times 1$
4. $567 = 5 \times 100 + 6 \times 10 + 7 \times 1$
5. $1,234 = 1 \times 1,000 + 2 \times 100 + 3 \times 10 + 4 \times 1$
6. $9,872 = 9 \times 1,000 + 8 \times 100 + 7 \times 10 + 2 \times 1$
7. $10,004 = 1 \times 10,000 + 4 \times 1$
8. $40,001 = 4 \times 10,000 + 1 \times 1$
9. $478,291 = 4 \times 100,000 + 7 \times 10,000 + 8 \times 1,000 + 2 \times 100 + 9 \times 10 + 1 \times 1$
10. $597,354 = 5 \times 100,000 + 9 \times 10,000 + 7 \times 1,000 + 3 \times 100 + 5 \times 10 + 4 \times 1$
11. $2 \times 10 + 1 \times 1 = 21$
12. $8 \times 10 + 3 \times 1 = 83$
13. $5 \times 100 + 6 \times 10 + 7 \times 1 = 567$
14. $2 \times 100 + 9 \times 10 + 8 \times 1 = 298$
15. $1 \times 10,000 + 0 \times 1,000 + 1 \times 100 + 0 \times 10 + 1 \times 1 = 10,101$
16. $2 \times 10,000 + 0, 1,000 + 0 \times 100 + 2 \times 10 + 0 \times 1 = 20,020$
17. $101 = 100 + 1$
18. $303 = 300 + 3$
19. $457 = 400 + 50 + 7$
20. $754 = 700 + 50 + 4$
21. $4,020 = 4,000 + 20$
22. $8,070 = 8,000 + 70$
23. $10,000 = 10,000$
24. $200,000 = 200,000$

2.2 Rounding Whole Numbers

1. 13 rounded to the nearest ten is 10.
2. 74 rounded to the nearest ten is 70.
3. 37 rounded to the nearest ten is 40.
4. 48 rounded to the nearest ten is 50
5. 55 rounded to the nearest ten is 60
6. 65 rounded to the nearest ten is 70
7. 101 rounded to the nearest hundred is 100.
8. 202 rounded to the nearest hundred is 200.
9. 109 rounded to the nearest hundred is 100.
10. 208 rounded to the nearest hundred is 200.
11. 625 rounded to the nearest hundred is 600.
12. 465 rounded to the nearest hundred is 500.
13. 1,999 rounded to the nearest hundred is 2,000.
14. 2,999 rounded to the nearest hundred is 3,000.
15. 903 rounded to the nearest thousand is 1,000.
    903 rounded to the nearest hundred is 900.
    903 rounded to the nearest ten is 900.
16. 901 rounded to the nearest thousand is 1,000.
    901 rounded to the nearest hundred is 900.
    901 rounded to the nearest ten is 900.
17. 9,084 rounded to the nearest ten thousand is 10,000.
    9,084 rounded to the nearest thousand is 9,000.
    9,084 rounded to the hearest hundred is 9,100.
    9,084 rounded to the nearest ten is 9,080.
18. 9,083 rounded to the nearest ten thousand is 10,000.
    9,083 rounded to the nearest thousand is 9,000.
    9,083 rounded to the nearest hundred is 9,100.
    9,083 rounded to the nearest ten is 9,080.

19. 479 rounded to the nearest hundred is 500.
479 rounded to the nearest ten is 480.

20. 489 rounded to the nearest hundred is 500.
489 rounded to the nearest ten is 490.

21. 1,005 rounded to the nearest thousand is 1,000.
1,005 rounded to the nearest hundred is 1,000.
1,005 rounded to the nearest ten is 1,010.

22. 3,105 rounded to the nearest thousand is 3,000.
3,105 rounded to the nearest hundred is 3,100.
3,105 rounded to the nearest ten is 3,110.

23. 10,654 rounded to the nearest ten thousand is 10,000.
10,654 rounded to the nearest thousand is 11,000.
10,654 rounded to the nearest hundred is 10,700.
10,654 rounded to the nearest ten is 10,650.

24. 20,745 rounded to the nearest ten thousand is 20,000.
20,745 rounded to the nearest thousand is 21,000.
20,745 rounded to the nearest hundred is 20,700.
20,745 rounded to the nearest ten is 20,750.

25. 1,153 miles rounded to the nearest hundred is 1,200.

26. 521 miles rounded to the nearest hundred mile is 500 miles.

27. 1,024 bytes rounded to the nearest thousand is 1,000.

28. Round 27 to 10s $27 \approx 30$
Vanessa should buy 5 packs.

29. Round 49,000 to 1,000s $49,000 \approx 50,000$
Juan should buy 4 bags.

2.3 Addition and Subtraction of Whole Numbers

1. $2 + 3 = 3 + 2$ demonstrates the Commutative Property of Addition.
2. $698 + 0 = 698$ demonstrates the Identity Property of Addition.
3. $28 + (17 + 52) = (28 + 17) + 52$ demonstrates the Associative Property of Addition.
4. $(9 + 10) + 11 = 11 + (9 + 10)$ demonstrates the Commutative Property of Addition.
5. $(15 + 0) + 8 = 15 + 8$ demonstrates the Identity Property of Addition.
6. The first string kicker scored: $6 + 7 + 11 + 8 = 32$ points.
7. The leading rusher ran for $16 + 56 + 80 + 110 = 262$ yards.
8. Sandy spends, $52 + 29 + 25 + 16 + 9 = \$131$.
9. Kyle spends, $245 + 35 + 25 + 48 + 19 = \$272$.
10. Catherine spends, $85 + 32 + 3 + 32 + 15 = \$167$.
11. Last year   21,729
this year $\underline{-20,982}$
                   747 fewer students this year
12. Balance $1,532
Withdrawal $\underline{-\$\ 255}$
New Balance $1,277
13. First hand   53 points
Second hand lost $\underline{-27}$ points
                             26 points

2.4 Multiplication and Division of Whole Numbers

1. $4(7 + 2) = 4(7) + 4(2) = 28 + 8 = 36$
2. $5(6 + 3) = 5(6) + 5(3) = 30 + 15 = 45$
3. $(20 + 6)3 = 20(3) + 6(3) = 60 + 18 = 78$
4. $(5 + 10)4 = 5(4) + 10(4) = 20 + 40 = 60$
5. $10 \div 5 = 2$, so $2(5) = 10$

4

6. $12 \div 6 = 2$, so $2(6) = 12$
7. $15 \div 5 = 3$, so $3(5) = 15$
8. $18 \div 9 = 2$, so $2(9) = 18$
9. $0 \div 10 = 0$, so $0(10) = 0$
10. $0 \div 8 = 0$, so $0(8) = 0$
11. $22 \div 0 =$ undefined
12. $36 \div 0 =$ undefined
13. $15 \times 31 = 31 \times 15$ demonstrates the Commutative Property for Multiplication.
14. $56 \times 1 = 56$ demonstrates the Identity Property for Multiplication.
15. $28 \times (17 \times 52) = (28 \times 17) \times 52$ demonstrates the Associate Property for Multiplication.
16. $(21 \times 8) \times 16 = (8 \times 21) \times 16$ demonstrates the Commutative Property of Multiplication.
17. $216 = 216 \times 1$ demonstates the Identity Property of Multiplication.
18. $57 = 3 \cdot 19$
19. 37 is prime
20. $39 = 3 \cdot 13$
21. $91 = 7 \cdot 13$
22. $123 = 3 \cdot 41$
23. 41 is prime
24. 97 is prime
25. 19 is prime
26. $267 = 3 \cdot 89$
27. $65 = 5 \cdot 13$
28. 89 is prime
29. $30 = 2 \cdot 3 \cdot 5$
30. 59 is prime
31. $51 = 3 \cdot 17$
32. $10 = 2 \cdot 5$
33. $50 = 2 \cdot 5 \cdot 5$
34. $99 = 3 \cdot 3 \cdot 11$
35. $200 = 2 \cdot 2 \cdot 2 \cdot 5 \cdot 5$
36. $6 = 2 \cdot 3$
37. $12 = 2 \cdot 2 \cdot 3$
38. $20 = 2 \cdot 2 \cdot 5$
39. $144 = 2 \cdot 2 \cdot 2 \cdot 2 \cdot 3 \cdot 3$
40. 547 is prime
41. $310 = 2 \cdot 5 \cdot 31$
42. $888 = 2 \cdot 2 \cdot 2 \cdot 3 \cdot 37$

43. $424 = 2 \cdot 2 \cdot 2 \cdot 53$
44. $900 = 2 \cdot 2 \cdot 3 \cdot 3 \cdot 5 \cdot 5$
45. $122 = 2 \cdot 61$
46. $133 = 7 \cdot 19$
47. GCF (9,16)   9 ,   16
    $3 \cdot 3$ $2 \cdot 2 \cdot 2 \cdot 2$
    GCF (9,16) = 1
48. GCF (16,18)   16 ,   18
    $(2) \cdot 2 \cdot 2 \cdot 2$ $(2) \cdot 3 \cdot 3$
    GCF (16,18) = 2
49. GCF (24,36)   24 ,   36
    $(2)(2) \cdot 2 \cdot 3$ $(2)(2)(3) \cdot 3$
    GCF (24,36) = 12
50. GCF (25,35)   25 ,   35
    $(5) \cdot 5$ $(5) \cdot 7$
    GCF (25,35) = 5
51. GCF (72,45)   72 ,   45
    $2 \cdot 2 \cdot 2 (3)(3)$ $(3)(3) \cdot 5$
    GCF (72,45) = 9
52. GCF (66,132)   66 ,   132
    $(2)(3)(11)$ $(2) \cdot 2 (3)(11)$
    GCF (66,1332) = 66
53. GCF (78,52)   78 ,   52
    $(2) \cdot 3 \cdot (13)$ $(2) \cdot 2 \cdot (13)$
    GCF (78,52) = 26
54. GCF (16,36,60)
    16 ,   36 ,   60
    $(2)(2) \cdot 2 \cdot 2$ $(2)(2) \cdot 3 \cdot 3$ $(2)(2) \cdot 3 \cdot 5$
    GCF (16,30,60) = 4
55. GCF (42,98,70)
    42 ,   98 ,   70
    $(2) \cdot 3 \cdot (7)$ $(2) \cdot 7 (7)$ $(2) \cdot 5 (7)$
    GCF (42,98,70) = 14
56. 153,000 households multiplied by \$6
    (153,000)(6) = \$918,000

5

57. 150 CDs multipled by $12
(150)(12) = $1,800

58. 18 miles for one gallon. For 16 gals.
multiply 18 by 16. (18)(16) = 288 miles

59. Saves $5 in one month. In 12 months
saves 5(12) = $60.

60. Divide $4,800,000 between three people
4,800,000 ÷ 3 = $1,600,000 each.

61. Traveled 1,150 miles in 23 hours. In
one hour traveled 1,150 ÷ 23 = 50 mph.

2.5 Exponents and Roots

1. $2^0 = 1$
2. $3^0 = 1$
3. $5^1 = 5$
4. $6^1 = 6$
5. $4^2 = 4 \cdot 4 = 16$
6. $3^2 = 3 \cdot 3 = 9$
7. $10^3 = 10 \cdot 10 \cdot 10 = 1,000$
8. $10^4 = 10 \cdot 10 \cdot 10 \cdot 10 = 10,000$
9. $2^3 = 2 \cdot 2 \cdot 2 = 8$
10. $3^3 = 3 \cdot 3 \cdot 3 = 27$
11. $3 \cdot 3 \cdot 3 \cdot 3 = 3^4$
12. $8 \cdot 8 \cdot 8 = 8^3$
13. $5 \cdot 5 \cdot 5 \cdot 5 = 5^4$
14. $2 \times 2 \times 2 \times 2 \times 2 \times 2 = 2^6$
15. $15^3 = 3,375$
16. $17^8 = 6,975,757,441$
17. $32^2 = 1,024$
18. $50^4 = 6,250,000$

19. $\sqrt{121} = 11$

20. $\sqrt{169} = 13$

21. $\sqrt[3]{343} = 7$

22. $\sqrt[3]{512} = 8$

23. $\sqrt{225} = 15$

24. $\sqrt{625} = 25$

25. 700 raised to the second power
$700^2 = 490,000$ people purchased life
insurance in 1990.

2.6 Order of Operations

1. $3 + 4 \cdot 6 = 3 + 24 = 27$
2. $6 \cdot 3 + 2 = 18 + 2 = 20$
3. $8 + 12 - 6 = 20 - 6 = 14$
4. $4 \cdot 2 + 3 \cdot 4 = 8 + 12 = 20$
5. $7 + 2 \cdot 8 - 4 = 7 + 16 - 4$
$= 23 - 4 = 19$
6. $6 \cdot 2 + 9 \cdot 8 = 12 + 72 = 84$
7. $2(4 + 7) = 2(11) = 22$
8. $(8 + 4)3 = (12)3 = 36$
9. $2[(3 + 6) \cdot 5] = 2[9 \cdot 5] = 2[45] = 90$
10. $7[(3 + 4) \cdot 6] = 7[7 \cdot 6] = 7[42] = 294$
11. $5 + 25 \div 5 = 5 + 5 = 10$
12. $7 + 18 \div 9 = 7 + 2 = 9$
13. $25 - 8(1 + 2) = 25 - 8(3) = 25 - 24 = 1$
14. $9 + 6(3 + 4) = 9 + 6(7) = 9 + 42 = 51$

15. $\dfrac{7(5)}{9-2} + 6 = \dfrac{35}{7} + 6 = 5 + 6 = 11$

16. $\dfrac{8(3+1)}{4(4)} + 15 = \dfrac{8(4)}{4(4)} + 15 =$

$\dfrac{32}{16} + 15 = 2 + 15 = 17$

17. $3 + 5(1 = 3^2) = 3 + 5(1 + 9) = 3 + 5(10)$
$= 3 + 50 = 53$
18. $2 + 7(2 + 4^2) = 2 + 7(2 + 16) = 2 + 7(18)$
$= 2 + 126 = 128$
19. $3 \cdot 2^2 - 1 + 4 = 3 \cdot 4 - 1 + 4$
$= 12 - 1 + 4 = 15$
20. $2 \cdot 3^2 - 7 + 5 = 2 \cdot 9 - 7 + 5$
$= 18 - 7 + 5 = 16$
21. $(7 + 1)^2 + 8 \div 4 = (8)^2 + 2 = 64 + 2 = 66$
22. $(2 + 6)^2 + 10 \div 2 = (8) + 5 = 64 + 5 = 69$
23. $5(2 + 3^3) + 7 = 5(2 + 27) + 7 = 5(29) + 7$
$= 145 + 7 = 152$

6

24. $4(5 + 2^3) + 3 = 4(5 + 8) + 3 = 4(13) + 3$
    $= 52 + 3 = 55$

25. $9 \times 14 + \{44 \div [19 - (5 + 3)]\}$
    $= 126 + \{44 \div [19 - (8)]\}$
    $= 126 + \{44 \div 11\} = 126 + 4 = 130$

26. $7 \times 15 + \{55 \div [23 - (9 + 3)]\}$
    $= 105 + \{55 \div [23 - 12]\}$
    $= 105 + \{55 \div 11\} = 105 + 5 = 110$

27. $84 \div 7 - \{3 \times [8 - (3 \times 2)]\}$
    $= 12 - \{3 \times [8 - 6]\} = 12 - \{3 \times 2\}$
    $= 12 - 6 = 6$

28. $62 \div 2 - \{5 \times [7 - (6 \div 3)]\}$
    $= 31 - \{5 \times [7 - 2]\} = 31 - \{5 \times 5\}$
    $= 31 - 25 = 6$

29. $5 \times \{(100 - 50 \div 10) - [(33 + 2) \div 5]$
    $\qquad + 4 \times 3\}$
    $= 5 \times \{(100 - 5) - \{35 \div 5\} + 4 \times 3\}$
    $= 5 \times \{95 - 7 + 4 \times 3\}$
    $= 5 \times \{95 - 7 + 12\} = 5 \times \{100\} = 500$

30. $7 \times \{(100 - 50 \div 10) - [(33 + 2) \div 5]$
    $\qquad + 4 \times 3\}$
    $= 7 \times \{(100 - 5) - \{35 \div 5\} + 12\}$
    $= 7 \times \{(100 - 5) - [35 \div 5] + 12\}$
    $= 7 \times \{95 - 7 + 12\} = 7 \times \{100\} = 700$

2.7 Averages and Estimation

1. $\dfrac{37 + 49}{2} = \dfrac{86}{2} = 43$

2. $\dfrac{43 + 89}{2} = \dfrac{132}{2} = 66$

3. $\dfrac{111 + 222 + 46 + 1}{4} = \dfrac{380}{4} = 95$

4. $\dfrac{360 + 945 + 524 + 867}{4} = \dfrac{2696}{4} = 674$

5. $\dfrac{999 + 2 + 52 + 101 + 46}{5} = \dfrac{1200}{5} = 240$

6. $\dfrac{788 + 1 + 47 + 64 + 100}{5} = \dfrac{1000}{5} = 200$

7. $\dfrac{1111 + 2222 + 3333 + 6666}{4}$
   $= \dfrac{13332}{4} = 3333$

8. $\dfrac{4444 + 5555 + 7777 + 8888}{4} =$
   $\dfrac{26664}{4} = 6666$

9. $\dfrac{12 + 581 + 0 + 75 + 17 + 587}{6} =$
   $\dfrac{1276}{6} = 212$

10. $\dfrac{35 + 782 + 43 + 0 + 246 + 16}{6} =$
    $\dfrac{1122}{6} = 187$

11. $74 + 22 + 58 + 90$
    $70 + 20 + 60 + 90 = 240$

12. $83 + 17 + 38 + 65$
    $80 + 20 + 40 + 70 = 210$

13. $235 + 117 + 89 + 22$
    $200 + 100 + 100 + 0 = 400$

14. $391 + 219 + 77 + 30$
    $400 + 200 + 100 + 0 = 700$

15. $10,000 - 2,500$
    $10,000 - 3,000 = 7,000$

16. $50,000 - 7,500$
    $50,000 - 8,000 = 42,000$

17. $25,659 + 31,743$
    $30,000 + 30,000 = 60,000$

18. $72,556 + 37,921$
    $70,000 + 40,000 = 110,000$

19. $123 + 456 + 789$
    $0 + 0 + 1,000 = 1,000$

20. $879 + 564 + 321$
    $1,000 + 1,000 + 0 = 2,000$

21. $\dfrac{3 + 7 + 2}{3} = \dfrac{12}{3} = 4$

    average of 4 passes per game

7

22. $\dfrac{3+5+100}{3} = \dfrac{108}{3} = 36$

average of 36 yards per game

23. $\dfrac{942+1098+1458+1830+2376+3660}{6}$

$= \dfrac{11364}{6} = \$1,894$

estimate

$\dfrac{1000+1000+1000+2000+2000+4000}{6}$

$= \dfrac{11000}{6} \approx \$1,900$

24. $\dfrac{28,500+30,258+31,580+22,490}{4}$

$= \dfrac{112,828}{4} = \$28,207$

estimate

$\dfrac{29,000+30,000+32,000+22,000}{4}$

$= \dfrac{113}{4} \approx \$28,000$

REVIEW

1.   See Section 2.1
2.   See Section 2.1
3.   See Section 2.1
4.   See Section 2.1
5.   See Section 2.1
6.   See Section 2.3
7.   See Section 2.4
8.   See Section 2.4
9.   See Section 2.5
10. See Section 2.4
11. See Section 2.4
12. See Section 2.4
13. See Section 2.3
14. See Section 2.3
15. See Section 2.3
16. See Section 2.4
17. See Section 2.4
18. See Section 2.4
19. See Section 2.4
20. See Section 2.4
21. See Section 2.6
22. See Section 2.7
23. See Section 2.2
24. See Section 2.3
25. See Section 2.4
26. See Section 2.4
27. $3 \times 100 + 1 \times 10 + 8 \times 1 = 318$
28. $5 \times 1,000 + 7 \times 100 + 9 \times 10 + 1 \times 1 = 5,791$
29. $9 \times 100 + 0 \times 10 + 0 \times 1 = 900$
30. $8,000 + 200 + 20 + 4 = 8,224$
31. $6 \times 1,000 + 0 \times 100 + 2 \times 10 + 7 \times 1 = 6,027$
32. $57 = 5 \times 10 + 7 \times 1$
33. $203 = 2 \times 100 + 0 \times 10 + 3 \times 1$
34. $8,502 = 8 \times 1,000 + 5 \times 100 + 0 \times 10 + 2 \times 1$
35. $18,511 = 1 \times 10,000 + 8 \times 1,000 + 5 \times 100 + 1 \times 10 + 1 \times 1$
36. $400,746 = 4 \times 100,000 + 7 \times 100 + 4 \times 10 + 6 \times 1$
37. $67 \approx 70$
38. $504 \approx 500$
39. $2,369 \approx 2,370$
40. $899 \approx 900$
41. $1,000,007 \approx 1,000,010$
42. $0 \div 8 = 0$
43. $8 \div 0 =$ undefined
44. $1^0 = 1$
45. $1^1 = 1$
46. $10^5 = 100,000$
47. $27^3 = 19,683$
48. 97 is prime.
49. 17 is prime.
50. 285; 5, 57
51. 300; 2, 3, 5
52. 41 is prime.
53. 72; 2, 3
54. 54; 3

55.   2 is prime.
56.   383 is prime.
57.   500 – 2, 5
58.   112 – 2, 7
59.   231 – 3, 7, 11
60.   GCF (26,40)   26   ,   40

                            ②· 13   ②· 2 · 2 · 5

     GCF(26,40) = 2

61.   GCF(50,110)   50   ,   110

                            ②⑤· 5   ②⑤· 11

     GCF(50,110) = 10

62.   GCF(18,25)   18   ,   25

                           2 · 3 · 3   5 · 5

     GCF(18,25) = 1

63.   GCF(8,20,44)   8 , 20 , 44

                          ②②· 2   ②②· 5   ②②· 11

     GCF(8,20,44) = 4

64. $\sqrt{64} = 8$

65. $\sqrt[3]{64} = 4$

66. $(2 + 7)(8 - 3) = (9)(5) = 45$
67. $2(6 - 4) + 7 = 2(2) + 7 = 4 + 7 = 11$
68. $(6 - 2)(1 + 5) = (4)(6) = 24$
69. $(2 + 3)^3 \div 5 = (5)^3 \div 5 = 125 \div 5 = 25$

70. $\dfrac{12+2}{8-1} + 5^2 = \dfrac{14}{7} + 25 = 2 + 25 = 27$

71. $[9 \div (2 + 1)] + 7 = [9 \div 3] + 7$
     $= 3 + 7 = 10$
72. $5^2 \div 6^2 = 25 + 36 = 61$
73. $9 + 7 \cdot 8 = 9 + 56 = 65$
74. $3(15 + 8) = 3(15) + 3(8) = 45 + 24 = 69$
75. $6 \cdot 6 \cdot 6 \cdot 6 = 6^4$

76.   {8,12,10} $\dfrac{8+12+10}{3} = \dfrac{30}{3} = 10$

77.   {27,33,24} $\dfrac{27+33+24}{3} = \dfrac{84}{3} = 28$

78.   {15,20,22,18,13,20}

     $\dfrac{15+20+22+18+13+20}{6} = \dfrac{108}{6} = 18$

79.   {101,336,267,212}

     $\dfrac{101+336+267+212}{4} = \dfrac{916}{4} = 229$

80.   {202,75,12,7}

     $\dfrac{202+75+12+7}{4} = \dfrac{296}{4} = 74$

81. $354 \approx 350$
     $729 \approx \underline{730}$
                1080

82. $267 \approx 300$
     $505 \approx \underline{500}$
               800

83. $385 \approx 390$
     $72 \approx - \underline{70}$
              320

84. $1467 \approx 1500$
     $983 \approx \underline{-1000}$
              500

85.   Erin's bills $100 + 68 + 25 + 75 + 20$
     $+ 60 + 75 = \$423$
     Her pay is $\$436 - \$423$ (her bills)
     She has $13 left, so even with the $70 in
     savings she cannot afford a $98 bike.

PRACTICE TEST

1. $3 \cdot 0 = 0$
   H. Zero Product Property

2. $17 + 14 = 14 + 17$
   C. Commutative Property of Addition

3. $18 \cdot 1 = 18$
   E. Identity for Multiplication

4. $2(3 + 19) = 2(3) + 2(19)$
   G. Distributive Property

5. $8 + (9 + 10) = (8 + 9) + 10$
   A. Associative Property for Adddition

6. $3 + 0 = 3$
   B. Identity Property for Addition

7. $2 \cdot 13 = 13 \cdot 2$
   F. Commutative Property for Multiplication

8. $7(10 \cdot 11) = (7 \cdot 10)11$
   D. Associative Property for Multiplication

9. $7,000 + 100 + 50 + 2 = 7,152$

10. $3 \times 100 + 0 \times 10 + 5 \times 1 = 305$

11. $10,704 = 1 \times 10,000 + 7 \times 100 + 4 \times 1$

12. 7,483 rounded to hundreths is 7,500.

13. 7,483 rounded to thousands is 7,000.

14. 7,483 rounded to tens is 7,480.

15. 7,483 rounded to ten thousands is 10,000

16. $\begin{aligned} 10,050 &\approx 10,100 \\ 4,149 &\approx -\underline{4,100} \\ & 6,000 \end{aligned}$

17. $15^1 = 15$

18. $6^3 = 216$

19. $160^4 = 655,360,000$

20. $17 + (8 - 2) = 17 + 6 = 23$

21. $(7^2 + 3) \div 4 = (49 + 3) \div 4 = 52 \div 4 = 13$

22. $2^4 + 4^2 = 16 + 16 = 32$

23. $2 + \{[5 - 1][6(3)] \cdot 2\} = 2 + \{(4)(18) \cdot 2\} = 2 + 144 = 146$

24. $6(5 + 4) = 6(5) + 6(4) = 30 + 24 = 54$

25. $5 \cdot 5 \cdot 5 = 5^3$

26. $2\ \underline{106}$
    $.\ 53 \qquad 106 = 2 \cdot 53$

27. Primes between 60 and 70 are 61, 67.

28. GCF(35,140) 35 , 140
    ⑤⑦ $2 \cdot 2$ ⑤⑦
    GCF(35,140) = 35

29. $\sqrt{144} = 12$

30. $\sqrt[3]{216} = 6$

31. $\dfrac{30+55+60+25+75}{5} = \dfrac{245}{5} = 49$

32. Apartment 1 will cost $300 per month.
    Apartment 2 will cost $100 rent + $40 + $26 for phone = $166.

    Income $625 - expenses $300 = $325.
    Apt. 1 leaves her $325 - $300 = $25
    Apt. 2 leaves her $325 - $166 = $159
    the first month and $325 - $126 = $199
    every other month. Apt. 2 seems a better
    arrangement financially.

CHAPTER 3: INTRODUCTION TO ALGEBRA

3.1 Simplifying Algebraic Expressions

1. $6x + 7y$; $6x$, $7y$
2. $3x^3 + 2x^2 + 6x + 7$; $3x^3$, $2x^2$, $6x$, 7
3. $4a + 5b + 6c + 7d$; $4a$, $5b$, $6c$, $7d$
4. $6xy + 5xy^2 + 4x + 6y$; $6xy$, $5xy^2$, $4x$, $6y$
5. $2a^2b + 3a^2 + 4b$; $2a^2b$, $3a^2$, $4b$
6. $9a + 7y$; $9a$, $7y$
7. $4xy + 7y + 8x$; $4xy$, $7y$, $8x$
8. $9d + 8e + 7g + 2z$; $9d$, $8e$, $7g$, $2z$
9. $9x^2 + 6x + 7$; $9x^2$, $6x$, 7
10. $4x^2 + 9x + 7y + 4$; $4x^2$, $9x$, $7y$, 4
11. The coefficient of x is 1.
12. The coefficient of 8xy is 8.
13. The coefficient of $19x^2y^2$ is 19.
14. The coefficient of z is 1.
15. The coefficient of 53y is 53.
16. The coefficient of 9 is 9.
17. The coefficient of 8ab is 8.
18. The coefficient of 3cd is 3.
19. The coefficient of $4x^2y^2z^2$ is 4.
20. The coefficient of 4fgh is 4.
21. $7(10a) = 70a$
22. $15(4x) = 60x$
23. $7y(6) = 42y$
24. $2x(27) = 54x$
25. $9(3ab) = 27ab$
26. $6(4x^2) = 24x^2$
27. $4x(x) = 4x^2$
28. $2(3y) = 6y$
29. $9(9xy) = 81xy$
30. $7(x^2) = 7x^2$
31. $9a(ab) = 9a^2b$
32. $4c(2d) = 8cd$
33. $7(x + 4) = 7x + 28$
34. $a(b + 6) = ab + 6a$
35. $(x + 4)7 = 7x + 28$
36. $3x(4x + 6y) = 12x^2 + 18xy$
37. $x(3 + y) = 3x + xy$
38. $2a(b - 9) = 2ab - 18a$
39. $8(4 - x) = 32 - 8x$
40. $(3a - 2)7 = 21a - 14$
41. $(3x + 4y)4 = 12x + 16y$
42. $9(8 + 2a) = 72 + 18a$
43. $9(a + 2b) = 9a + 18b$
44. $7(2c + d) = 14c + 7d$
45. $(3a + 2b)a = 3a^2 + 2ab$
46. $(3z + y)2a = 6az + 2ay$
47. $3b(c + 2d) = 3bc + 6bd$
48. $4x(x + 2y) = 4x^2 + 8xy$
49. $3z(2x + 3y) = 6xz + 9yz$
50. $17(4a + 8b) = 68a + 136b$
51. $15(9z + 7x) = 135z + 105x$
52. $(8a + 6b)14 = 112a + 84b$
53. $2x + 3x + 10x - 3x = 12x$
54. $6a - 5a + 10a = 11a$
55. $12y + y - 11y = 2y$
56. $4xy + 5xy - xy = 8xy$
57. $3a^2 + 4a + 2a - a = 3a^2 + 5a$
58. $2y^2 + 6y + 4xy = 2y^2 + 6y + 4xy$
59. $3x^2 + 7x^2 - 2x^2 + 5x + 2x + x = 8x^2 + 8x$
60. $2y^2 + 4y^2 + 6xy + 3x^2 + 2x^2$
    $= 6y^2 + 6xy + 5x^2$
61. $18t + 23t + t = 42t$
62. $b + 6b + 19b + 21b = 47b$
63. $4a^2 + 3a^2 + 5a + 7a = 7a^2 + 12a$
64. $9b + 7b + 7b + 4b = 27b$
65. $19x^2 + 9x + 9x^2 + 14x = 28x^2 + 23x$
66. $5z^2 + 9y + 12z^2 = 17z^2 + 9y$
67. $15c + 5cd + 8cd = 15c + 13cd$
68. $2m + 9m + 8m + 4m^2 = 19m + 4m^2$
69. $16jk + 9j + 14k + 4jk = 20jk + 9j + 14k$
70. $13x + 19y + 14z = 13x + 19y + 14z$
71. $23x + 14x + 9x + 2x = 48x$
72. $15x + 7y + 8x + 12y = 23x + 19y$
73. $4(a + 6) + 6$
    $= 4a + 24 + 6$
    $= 4a + 30$
74. $y + 6y + 3(y + 4)$
    $= y + 6y + 3y + 12$
    $= 10y + 12$

75. $4 + 3(x + 7)$
$= 4 + 3x + 21$
$= 3x + 25$

76. $4(x + 3) + 8(x + 6)$
$= 4x + 12 + 8x + 48$
$= 12x + 60$

77. $6 + (x + 3)3$
$= 6 + 3x + 9$
$= 3x + 15$

78. $3 + 4(x + 7) + 6$
$= 3 + 4x + 28 + 6$
$= 4x + 37$

79. $7x + 6y + 3(x + 4)$
$= 7x + 6y + 3x + 12$
$= 10x + 6y + 12$

80. $12(x + y) + 4y + 3x$
$= 12x + 12y + 4y + 3x$
$= 15x + 16y$

81. $2x + 7(3x + 1) + 6x$
$= 2x + 21x + 7 + 6x$
$= 29x + 7$

82. $a(7 + b) + 3ab$
$= 7a + ab + 3ab$
$= 7a + 4ab$

83. $4a(b + a) + 3b(a + 4) + 7(b)$
$= 4ab + 4a^2 + 3ab + 12b + 7b$
$= 4a^2 + 7ab + 19b$

84. $3x(x + 3y + 9) + 7x(9)$
$= 3x^2 + 9xy + 27x + 63x$
$= 3x^2 + 9xy + 90x$

85. $19(a + 3b + c) + 8c(4)$
$= 19a + 57b + 19c + 32c$
$= 19a + 57b + 51c$

86. $13(3j + 2k) + 8jk$
$= 39j + 26k + 8jk$

87. $9c(8c + 5d) + 3(4d + 5)$
$= 72c^2 + 45cd + 12d + 15$

88. $8(3x + 2x) + 9(3x + y)$
$= 24x + 16x + 27x + 9y$
$= 67x + 9y$

89. $(8a + 3b)4 + 7(a + 4b)$
$= 32a + 12b + 7a + 28b$
$= 39a + 40b$

90. $5(2m + 3n) + 6(8n + 4m)$
$= 10m + 15n + 48n + 24m$
$= 34m + 63n$

91. $7x(x + 3) + 2x(x + 4)$
$= 7x^2 + 21x + 2x^2 + 8x$
$= 9x^2 + 29x$

92. $9y(8 + y) + 5y(3y + 8)$
$= 72y + 9y^2 + 15y^2 + 40y$
$= 24y^2 + 112y$

3.2 Exponents

1. $3 \cdot 3 = 3^2$
2. $4 \cdot 4 \cdot 4 = 4^3$
3. $9 \cdot 9 \cdot 9 \cdot 9 = 9^4$
4. $x \cdot x \cdot x \cdot x \cdot x = x^5$
5. $7 \cdot 7 \cdot 7 \cdot 7 = 7^4$
6. $23 \cdot 23 \cdot 23 \cdot 23 \cdot 23 = 23^5$
7. $4y \cdot 4y \cdot 4y \cdot 4y = (4y)^4$
8. $19 \cdot 19 \cdot 19 \cdot 19 \cdot 19 \cdot 19 = 19^6$
9. $ab \cdot ab \cdot ab = (ab)^3$
10. $3pg \cdot 3pg = (3pg)^2$
11. $9 = 3^2$
12. $16 = 4^2$ or 2^4
13. $81 = 9^2$ or 3^4
14. $32 = 2^5$
15. $125 = 5^3$
16. $3^0 = 1$
17. $2^0 = 1$
18. $15^1 = 15$
19. $27^1 = 27$
20. $1^7 = 1 \cdot 1 \cdot 1 \cdot 1 \cdot 1 \cdot 1 \cdot 1 = 1$
21. $1^5 = 1 \cdot 1 \cdot 1 \cdot 1 \cdot 1 = 1$
22. $5^4 = 5 \cdot 5 \cdot 5 \cdot 5 = 625$
23. $3^6 = 3 \cdot 3 \cdot 3 \cdot 3 \cdot 3 \cdot 3 = 729$
24. $(4x)^2 = 4x \cdot 4x = 16x^2$
25. $(5y)^3 = 5y \cdot 5y \cdot 5y = 125y^3$
26. $9^3 = 9 \cdot 9 \cdot 9 = 729$
27. $10^4 = 10 \cdot 10 \cdot 10 \cdot 10 = 10,000$
28. $1^{45} = 1$ multiplied by itself 45 times $= 1$
29. $7^3 = 7 \cdot 7 \cdot 7 = 343$
30. $6^5 = 6 \cdot 6 \cdot 6 \cdot 6 \cdot 6 = 7,776$
31. $a^2 \cdot a^6 = a^{2+6} = a^8$

32. $7^2 \cdot 7^3 = 7^{2+3} = 7^5$
33. $x^5 \cdot x = x^{5+1} = x^6$
34. $4^2 \cdot 4^3 = 4^{2+3} = 4^5$
35. $12^2 \cdot 12^2 = 12^{2+2} = 12^4$
36. $y^4 \cdot y^7 = y^{4+7} = y^{11}$
37. $b \cdot b = b^{1+1} = b^2$
38. $c^6 \cdot c^7 = c^{6+7} = c^{13}$
39. $z \cdot z^2 = c^{1+2} = z^3$
40. $(2^2)^6 = 2^{2 \cdot 6} = 2^{12}$
41. $(5^2)^3 = 5^{2 \cdot 3} = 5^6$
42. $(1^9)^{16} = 1^{9 \cdot 16} = 1^{144}$
43. $(a^8)^5 = a^{8 \cdot 5} = a^{40}$
44. $(x^3)^4 = x^{3 \cdot 4} = x^{12}$
45. $(x^7)^7 = x^{7 \cdot 7} = x^{49}$
46. $(2y^2)^6 = 2^{1 \cdot 6} \cdot y^{2 \cdot 6} = 2^6 \cdot y^{12} = 64y^{12}$
47. $(z^5)^4 = z^{5 \cdot 4} = z^{20}$
48. $(d^4)^7 = d^{4 \cdot 7} = d^{28}$
49. $(b^3)^5 = b^{3 \cdot 5} = b^{15}$
50. $(ab)^2 = a^{1 \cdot 2} \cdot b^{1 \cdot 2} = a^2b^2$
51. $(xy)^5 = x^{1 \cdot 5} \cdot y^{1 \cdot 5} = x^5y^5$
52. $(fg)^6 = f^{1 \cdot 6} \cdot g^{1 \cdot 6} = f^6g^6$
53. $(3a)^8 = 3^{1 \cdot 8} \cdot a^{1 \cdot 8} = 3^8 \cdot a^8 = 6,561a^8$
54. $(4x)^4 = 4^{1 \cdot 4} \cdot x^{1 \cdot 4} = 4^4 \cdot x^4 = 256x^4$
55. $(7ab)^2 = 7^{1 \cdot 2} \cdot a^{1 \cdot 2} \cdot b^{1 \cdot 2} = 7^2a^2b^2$
    $= 49a^2b^2$
56. $(2xy)^5 = 2^{1 \cdot 5} \cdot x^{1 \cdot 5} \cdot y^{1 \cdot 5} = 2^5x^5y^5$
    $= 32x^5y^5$
57. $(xyz)^4 = x^{1 \cdot 4} \cdot y^{1 \cdot 4} \cdot z^{1 \cdot 4} = x^4y^4z^4$
58. $(6pq)^2 = 6^{1 \cdot 2} \cdot p^{1 \cdot 2} \cdot q^{1 \cdot 2} = 6^2p^2q^2$
    $= 36p^2q^2$
59. $(9x)^3 = 9^{1 \cdot 3} \cdot x^{1 \cdot 3} = 9^3 \cdot x^3 = 729x^3$
60. $(5y)^5 = 5^{1 \cdot 5} \cdot y^{1 \cdot 5} = 5^5 \cdot y^5 = 3125y^5$
61. $(a^3)^7 = a^{3 \cdot 7} = a^{21}$
62. $x^4 \cdot x^7 = x^{4+7} = x^{11}$
63. $(y^8)^7 = y^{8 \cdot 7} = y^{56}$
64. $(jk)^{23} = j^{1 \cdot 23} \cdot k^{1 \cdot 23} = j^{23}k^{23}$
65. $s^8 \cdot s^7 = s^{8+7} = s^{15}$
66. $(x^{12})^{10} = x^{12 \cdot 10} = x^{120}$
67. $(x^2y^4)^4 = x^{2 \cdot 4} \cdot y^{4 \cdot 4} = x^8y^{16}$
68. $(y^6 \cdot y^4)^5 = (y^{6+4})^5 = (y^{10})^5 = y^{10 \cdot 5} = y^{50}$
69. $(a^4b^6)^9 = a^{4 \cdot 9} \cdot b^{6 \cdot 9} = a^{36}b^{54}$
70. $(b^4 \cdot b^8)^{11} = (b^{4+8})^{11} = (b^{12})^{11}$
    $= b^{12 \cdot 11} = b^{132}$

3.3 Evaluating Algebraic Expressions

1. $2a + 6$, if $a = 3$
   $= 2(a) + 6 = 2(3) + 6 = 6 + 6 = 12$
2. $7x - 8$, if $x = 4$
   $= 7(x) - 8 = 7(4) - 8 = 28 - 8 = 20$
3. $4a + 9$, if $a = 5$
   $= 4(a) + 9 = 4(5) + 9 = 20 + 9 = 29$
4. $9x + 9$, if $x = 9$
   $= 9(x) + 9 = 9(9) + 9 = 81 + 9 = 90$
5. $7b + 2$, if $b = 3$
   $= 7(b) + 2 = 7(3) + 2 = 21 + 2 = 23$
6. $10y - 5$, if $y = 2$
   $= 10(y) - 5 = 10(2) - 5 = 20 - 5 = 15$
7. $2a + 6$, if $a = 0$
   $= 2(a) + 6 = 2(0) + 6 = 0 + 6 = 6$
8. $7x - 8$, if $x = 8$
   $= 7(x) - 8 = 7(8) - 8 = 56 - 8 = 48$
9. $4a + 9$, if $a = 10$
   $= 4(a) + 9 = 4(10) + 9 = 40 + 9 = 49$
10. $9x + 9$, if $x = 1$
    $= 9(x) + 9 = 9(1) + 9 = 9 + 9 = 18$
11. $7b + 2$, if $b = 0$
    $= 7(b) + 2 = 7(0) + 2 = 0 + 2 = 2$
12. $10y - 5$, if $y = 12$
    $= 10(y) - 5 = 10(12) - 5$
    $= 120 - 5 = 115$
13. $x^2 + 2x - 4$, if $x = 4$
    $= (x)(x) + 2(x) - 4 = (4)(4) + 2(4) - 4$
    $= 16 + 8 - 4 = 20$
14. $7x + 9$, if $x = 6$
    $= 7(6) + 9 = 42 + 9 = 51$
15. $9y - 6$, if $y = 2$
    $= 9(2) - 6 = 18 - 6 = 12$
16. $3a + 4$, if $a = 0$
    $= 3(0) + 4 = 0 + 4 = 4$
17. $2m + 5$, if $m = 9$
    $= 2(9) + 5 = 18 + 5 = 23$
18. $q + 9$, if $q = 17$
    $= 17 + 9 = 26$
19. $29 + r$, if $r = 8$
    $= 29 + 8 = 37$
20. $t^2 + 2t + 4$, if $t = 1$
    $= (1)^2 + 2(1) + 4 = 1 + 2 + 4 = 7$

21. $a^2 + 2a + 4$, if $a = 8$
$= (8)^2 + 2(8) + 4 = 64 + 16 + 4 = 84$

22. $j^2 + 6j + 4$, if $j = 7$
$= (7)^2 + 6(7) + 4 = 49 + 42 + 4 = 95$

23. $x^2 + 8$, if $x = 6$
$= (6)^2 + 8 = 36 + 8 = 44$

24. $y^2 + y$, if $y = 5$
$= (5)^2 + 5 = 25 + 5 = 30$

25. $2m^2 + m + 9$, if $m = 0$
$= 2(0)^2 + 0 + 9 = 0 + 0 + 9 = 9$

26. $3x^2 + 4x + 7$, if $x = 3$
$= 3(3)^2 + 4(3) + 7 = 27 + 12 + 7 = 46$

27. $x + xy + y$, if $x = 4$ and $y = 6$
$= 4 + (4)(6) + 6 = 4 + 24 + 6 = 34$

28. $5a + ab + 4b$, if $a = 5$ and $b = 9$
$= 5(5) + (5)(9) + 4(9)$
$= 25 + 45 + 36 = 106$

29. $7a^2 + 9a - 7$, if $a = 8$
$= 7(8)^2 + 9(8) - 7 = 448 + 72 - 7 = 513$

30. $4x^3 + 6x^2 + 9x + 5$, if $x = 4$
$= 4(4)^3 + 6(4)^2 + 9(4) + 5$
$= 256 + 96 + 36 + 5 = 393$

31. $5x + 7xy + 7y$, if $x = 6$ and $y = 7$
$= 5(6) + 7(6)(7) + 7(7)$
$= 30 + 294 + 49 = 373$

32. $6x + 7xy + 4y$, if $x = 3$ and $y = 4$
$= 6(3) + 7(3)(4) + 4(4)$
$= 18 + 84 + 16 = 118$

33. $5a^3 + 2a^2 + a + 6$, if $a = 9$
$= 5(9)^3 + 2(9)^2 + 9 + 6$
$= 3,645 + 162 + 9 + 6 = 3,822$

34. $3x^2 + 4x + 6$, if $x = 8$
$= 3(8)^2 + 4(8) + 6 = 192 + 32 + 6 = 230$

35. $9x^3 + 16x^2 + 4x$, if $x = 0$
$= 9(0)^3 + 16(0)^2 + 4(0) = 0 + 0 + 0 = 0$

36. $3m^2 + 2n^2 + 6n + 9$, if $m = 3$ and $n = 4$
$= 3(3)^2 + 2(4)^2 + 6(4) + 9$
$= 27 + 32 + 24 + 9 = 92$

37. $7j + 9jk + 6k$, if $j = 2$ and $k = 7$
$= 7(2) + 9(2)(7) + 6(7)$
$= 14 + 126 + 42 = 182$

38. $3c^2 + 2d + 5cd$, if $c = 1$ and $d = 0$
$= 3(1)^2 + 2(0) + 5(1)(0) = 3 + 0 + 0 = 3$

39. $5x + 7y + 9xy$, if $x = 3$ and $y = 8$
$= 5(3) + 7(8) + 9(3)(8)$
$= 15 + 56 + 216 = 287$

40. $12a + 16b + 12ab + 3a^2$, if $a = 1$
and $b = 3$
$= 12(1) + 16(3) + 12(1)(3) + 3(1)^2$
$= 12 + 48 + 36 + 3 = 99$

3.4 Translating Words Into Algebraic Expressions

1. $14 \div x$ or $\dfrac{14}{x}$

2. $17x$

3. $x + 20$

4. $x - 9$

5. $4 - x$

6. $3x$

7. $8 + x$ or $x + 8$

8. $21 - x$

9. $x \div 2$ or $\dfrac{x}{2}$

10. $6x$

11. $13 - x$

12. $28 + x$ or $x + 28$

13. $1 - x$

14. $x - 1$

15. $17 \div x$ or $\dfrac{17}{x}$

16. $x - 24$

17. $2x$

18. $x - 50$

19. $10x$

20. $50 - x$

21. $2x$

22. $34 + x$ or $x + 34$

23. $x \div 25$ or $\dfrac{x}{25}$

24. $45 - x$

25. $6x$

14

26.   x − 76

27.   67 + x or x + 67

28.   54x

29.   x + 43 or 43 + x

30. $76 \div x$ or $\dfrac{76}{x}$

31.   19 − 9

32.   x − 10

33.   x + 10

34.   2x

35.   x + 10

36.   x − 20

37.   x − 19

38.   x − 9

39.   8x

40.   v + a

41.   j(f)

42.   d + b

43.   10x

44. $\dfrac{x}{100}$

45. $\dfrac{x}{7}$, 6x, 52x

46.   10x

47. $\dfrac{150}{x}$

48.   11 − x

49.   30 − x − y

50.   50 + x − y

51.   10x

52.   700 − x − y

53.   x + 6 or 6 + x

54.   x − 3

55.   2x + x = 3x

56.   27 − x + y

57.   2x

58.   x − 4

REVIEW

1.   See Section 3.1

2.   See Section 3.1

3.   See Section 3.1

4.   See Section 3.1

5.   See Section 3.1

6.   See Section 3.1

7.   See Section 3.1

8.   See Section 3.1

9.   See Section 3.2

10.   See Section 3.2

11.   3x + 2y + 92; 3x, 2y, 92

12. $4x^2 + 5x + 7$; $4x^2$, 5x, 7

13.   6a + 4ab + 7b; 6a, 4ab, 7b

14.   4x + 6y; 4x, 6y

15.   9a + 7; 9a, 7

16.   14b + 3a; 14b, 3a

17.   2c + 3d + 4cd; 2c, 3d, 4cd

18.   9x + 7y + 46; 9x, 7y, 46

19.   3jk + 2j + 7k + 6; 3jk, 2j, 7k, 6

20.   The variables are x^3, x^2, x, xy, xy, y^2, y.

21.   The variables are a^2, a, ab, b^2, b.

22.   The variables are c^2, c, cd, d.

23.   The coefficients are 6, 4, 5, 6, 1, 3, 1, 8.

24.   The coefficients are 3, 4, 6, 7, 1, 9.

25.   The coefficients are 4, 1, 1, 5, 7.

26.   The constant terms are 9, 46, 86.

27.   The constant terms are 19, 2.

28.   The constant terms are 15, 0.

29.   6x + x + 9x = 16x

30.   3y + 18y − y + 4x − 3x = 20y + x

31.   2a + 4a + ab + 3ab + 6b − b
       = 6a + 4ab + 5b

32.   8y − 2y = 6y

33. $3x^2 + 5x + 2x^2 + 6x = 5x^2 + 11x$

34.   3x + 2x + 7x = 12x

35. $7x^2 + 9x + 3x^2 + 4x^2 = 14x^2 + 9x$

36.   ab + 3a + 7b

37. $y + 3y + 9y + 4y^2 = 13y + 4y^2$

38. $3m^2 + 6m^2 + 9m + 5m = 9m^2 + 14m$

39.   7c + 9d + 3cd

40. $4x^2 + 0 + 9x^2 + 6 = 13x^2 + 6$

41.   4x + x + 2xy = 5x + 2xy

15

42. $3ab + 6ab + 96a = 9ab + 96a$

43. $7x + 9x + 3y + 9y + 6xy + 2xy$
$= 16x + 12y + 8xy$

44. $7(x + 3) = 7(x) + 7(3) = 7x + 21$

45. $(a + 5)8 = a(8) + 5(8) = 8a + 40$

46. $(y - 8)9y = y(9y) - 8(9y) = 9y^2 - 72y$

47. $x(y - 1) = x(y) - x(1) = xy - x$

48. $7(a + b) = 7(a) + 7(b) = 7a + 7b$

49. $3(2x + 4) = 3(2x) + 3(4) = 6x + 12$

50. $(2y + 9)4 = 2y(4) + 9(4) = 8y + 36$

51. $4x(3 + y) = 4x(3) + 4x(y) = 12x + 4xy$

52. $5(x + 4) = 5(x) + 5(4) = 5x + 20$

53. $6(y - 3) = 6(y) - 6(3) = 6y - 18$

54. $4 + 6(3 + x) = 4 + 18 + 6x = 22 + 6x$

55. $x + 3(x + 9) = x + 3x + 27 = 4x + 27$

56. $(y + 4)5 + 3 = 5y + 20 + 3 = 5y + 23$

57. $2(a + 4) + 3(2 + a) = 2a + 8 + 6 + 3a$
$= 5a + 14$

58. $7 + 3(x + 6) = 7 + 3x + 18 = 3x + 25$

59. $y + 4(y + 3) = y + 4y + 12 = 5y + 12$

60. $(2a + 4)5 + 6 = 10a + 20 + 6 = 10a + 26$

61. $(2c + d)3 + 2d = 6c + 3d + 2d = 6c + 5d$

62. $4(2a + 2b) + 3(a + b)$
$= 8a + 8b + 3a + 3b = 11a + 11b$

63. $9(2x + 3y) + (3x + 4y)6$
$= 18x + 27y + 18x + 24y = 36x + 51y$

64. $7^2 = 49$

65. $5^3 = 125$

66. $x^9 = x^9$

67. $93^0 = 1$

68. $1^{56} = 1$

69. $a^5 \cdot a^7 = a^{5+7} = a^{12}$

70. $(x^6)^9 = x^{6 \cdot 9} = x^{54}$

71. $(gh)^6 = g^6h^6$

72. $(y^2 \cdot y^4)^5 = (y^{2+4})^5 = (y^6)^5 = y^{30}$

73. $(y^4z^2)^4 = y^{4 \cdot 4} \cdot z^{2 \cdot 4} = y^{16}z^8$

74. $(4x^5)^3 = 4^{1 \cdot 3} \cdot x^{5 \cdot 3} = 4^3 \cdot x^{15} = 64x^{15}$

75. $3x + 6$, if $x = 4$
$= 3(x) + 6 = 3(4) + 6 = 12 + 6 = 18$

76. $7a + 3$, if $a = 6$
$= 7(b) + 3 = 42 + 3 = 45$

77. $2xy + x^2$, if $x = 3$ and $y = 4$
$= 2(3)(4) + (3)^2 = 24 + 9 = 33$

78. $3x^2 + 2x + 4$, if $x = 0$
$= 3(0)^2 + 2(0) + 4 = 0 + 0 + 4 = 4$

79. $2a + b$, if $a = 3$
$= 2(3) + 6 = 6 + 6 = 12$

80. $4b + 9$, if $b = 5$
$= 4(5) + 9 = 20 + 9 = 29$

81. $7x + 6$, if $x = 0$
$= 7(0) + 6 = 0 + 6 = 6$

82. $9y + 1$, if $y = 1$
$= 9(1) + 1 = 9 + 1 = 10$

83. $8z + z$, if $z = 8$
$= 8(8) + 8 = 64 + 8 = 72$

84. $4x + 5y + 9$, if $x = 2$ and $y = 3$
$= 4(2) + 5(3) + 9 = 8 + 15 + 9 = 32$

85. $9c + 3cd + 4$, if $c = 4$ and $d = 4$
$= 9(4) + 3(4)(4) + 4 = 36 + 48 + 4 = 88$

86. $2m^2 + 3m + 9$, if $m = 5$
$= 2(5)(5) + 3(5) + 9 = 50 + 15 + 9 = 74$

87. $2m + 3mn + 7n$, if $m = 6$ and $n = 2$
$= 2(6) + 3(6)(2) + 7(2)$
$= 12 + 36 + 14 = 62$

88. $3x^2 + 6x + 19$, if $x = 7$
$= 3(7)(7) + 6(7) + 19$
$= 147 + 42 + 19 = 208$

89. $4y^2 + 4y + 16$, if $y = 8$
$= 4(8)(8) + 4(8) + 16$
$= 256 + 32 + 16 = 304$

90. $x + 7$ or $7 + x$

91. $x - y$

92. $2x$

93. $9 \div x$ or $\dfrac{9}{x}$

94. $x + y$ or $y + x$

95. $72x$

96. $15 \div x$ or $\dfrac{15}{x}$

97. $19 - x$

98. $6 + x$ or $x + 6$

99. $14 - 7$

100. $x - 9$

101. $x - y$

102. $3 - 7$

103. $x - 7$

104. $9 + 10$ or $10 + 9$

105. $x - 21$

16

106. $3x$

107. $x \div y$ or $\dfrac{x}{y}$

108. $x + 19$ or $19 + x$

109. $36 \div x$ or $\dfrac{36}{x}$

110. $x - 9$
111. $2x$
112. $50 + x$
113. $43 - x + y$

PRACTICE TEST

1. $6x + 9xy + 7y$; $6x$, $9xy$, $7y$

2. y The coefficient is 1.

3. $7y$ The coefficient is 7.

4. $81xy$ The coefficient is 81.

5. $27a$ The coefficient is 27.

6. $10a - 4a = 6a$

7. $5y - y = 4y$

8. $4ab + 5ab = 9ab$

9. $3x + 8x + 7y + 4x + x - y$
$= 16x + 6y$

10. $4x^3 + 3x^2 + 2x^2 + 5x + x + 9 - 2$
$= 4x^3 + 5x^2 + 6x + 7$

11. $4(x + 7) = 4x + 28$

12. $(3 + x)\,9x = 27x + 9x^2$

13. $(x - 8)5 = 5x - 40$

14. $3y(2y - 4) = 6y^2 - 12y$

15. $6(4x + 3) - 10 + x = 24x + 18 - 10 + x$
$= 25x + 8$

16. $(9 + 4)6 + 2(3 + a) = 6a + 24 + 6 + 2a$
$= 8a + 30$

17. $3x + 4x + 7(x + 9) = 3x + 4x + 7x + 63$
$= 14x + 63$

18. $8x + 1$, if $x = 6$
$= 8(6) + 1 = 48 + 1 = 49$

19. $3x^2 + 5x + 10$, if $x = 9$
$= 3(9)^2 + 5(9) + 10$
$= 243 + 45 + 10 = 298$

20. $3x^2 + 4x + 9$, if $x = 0$
$= 3(0)^2 + 4(0) + 9 = 0 + 0 + 9 = 9$

21. $x + 10$ or $10 + x$

22. $17 - x$

23. xy or $x \cdot y$

24. $x \div y$ or $\dfrac{x}{y}$

25. $x \div 4$ or $\dfrac{x}{4}$

26. $6 - x$

27. $8x$

28. $x - 6$

29. $x \div 7$ or $\dfrac{x}{7}$

30. $23 + x$ or $x + 23$

31. $(7x^2)^3 = 7^3x^6 = 343x^6$

32. $a^4 \cdot a^{17} = a^{4+17} = a^{21}$

17

CUMULATIVE REVIEW OF CHAPTERS 1, 2, AND 3

1. 3 x 4 = 4 x 3 is an example of the Commutative Property of Multiplication.

2. The symbol "⊆" is used to represent that one set is a subset of another set.

3. The Associative Property of Multiplication states that, for all a, b, c ∈ Reals, $(a \cdot b) \cdot c = a \cdot (b \cdot c)$

4. The Commutative Property of Addition states that, for all a, b ∈ Reals, $a + b = b + a$.

5. The Distributive Property states that, for all a, b, c ∈ Reals, $a(b + c) = a \cdot b + a \cdot c$.

6. The set of Natural Numbers = {1,2,3 . . .}.

7. The set of Whole Numbers = {0,1,2,3 . . .}.

8. In the expression 35x, "x" is a variable.

9. A whole number greater than 1 whose only factors are itself and one is a prime number.

10. In the expression $3y^2 + 7$, the number "7" is a constant.

11. $15(x + 2y) = 15(x) + 15(2y) = 15x + 30y$

12. $4(a + 17) + 12 = 4(a) + 4(17) + 12 = 4a + 68 + 12 = 4a + 80$

13. $31j - 5j = 26j$

14. $6p + 2r + 7p - r = 13p + r$

15. $5 \cdot 5 \cdot 5 = 5^3$

16. 76,082 rounded to the nearest ten is 76,080.

17. 76,082 rounded to the nearest hundred is 76,100.

18. 76,082 rounded to the nearest thousand is 76,000.

19. 76,082 rounded to the nearest ten thousand is 80,000.

20. $(12)5 = 60$

21. $\sqrt{81} = 9$

22. $6^3 = 216$

23. $5 \cdot 0 = 0$

24. $7^0 = 1$

25. $8 + 2 \cdot 13 = 8 + 26 = 34$

26. $34 + \sqrt{9} = 34 + 3 = 37$

27. $\dfrac{16}{2 + 2} = \dfrac{16}{4} = 4$

28. $\dfrac{41}{0} = undefined$

29. $(x^2 \cdot x^4)^5 = (x^6)^5 = x^{30}$

30. $(3y^4)^3 = 3^3 y^{12} = 27y^{12}$

31. $\sqrt[3]{64} = 4$

32. $10ab + 7ab - ab = 16ab$

18

33. $5(3x + 4) - 10 + x = 15x + 20 - 10 + x$
    $= 16x + 10$

34. $5 \cdot 6 + 45 \div 9 = 30 + 5 = 35$

35. $28 \cdot 7 - 6 \cdot 8 = 196 - 48 = 148$

36. $31 - [2(4 + 5) + 8] = 31 - [2(9) + 8]$
    $= 31 - [18 + 8] = 31 - 26 = 5$

37. $\{24 - [5 - (4 - 2 \div 2)\} \div (3 + 16 \div 2)$
    $= \{24 - [5 - (4 - 1)]\} \div (3 + 8)$
    $= \{24 - [5 - 3]\} \div 11$
    $= \{24 - 2\} \div 11 = 22 \div 11 = 2$

38. $9{,}782 \approx 9{,}800$
    $2{,}934 \approx + \underline{2{,}900}$
    $\phantom{2{,}934 \approx +} 12{,}700$

39. $4y^2 - y + 3$ when $y = 3$
    $= 4(y)(y) - (y) + 3$
    $= 4(3)(3) - (3) + 3 = 36 - 3 + 3 = 36$

40. $\dfrac{7 - h}{3 + h}$ when $h = 7$

    $= \dfrac{7 - 7}{3 + 7} = \dfrac{0}{10} = 0$

41. The sum of 2L and 2W = 2L + 2W.

42. Forty less than three times R is 3R − 40.

43. {x,y,z}, {x,y}, {x,z}, {y,z}, {x}, {y},
    {z}, {}

44.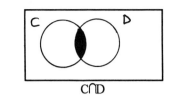
C∩D

45.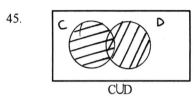
C∪D

19

CHAPTER 4: INTEGERS

4.1 The Set of Integers

1. 4 is greater than 2
2. −6 is greater than −16
3. 4 is less than 8
4. 4 is greater than −8
5. 0 is greater than −4
6. 47 is less than 49
7. −43 is greater than −44
8. −194 is less than 94
9. 14 is greater than 0
10. 5 is greater than −9
11. $9 > 5$
12. $-9 < -2$
13. $15 < 17$
14. $19 > -4$
15. $0 > -4$
16. $-9 < 0$
17. $-64 < -63$
18. $1 < 10$
19. $-1 > -10$
20. $1 > -10$
21. $7 < 16$
22. $29 > -30$
23. $-7 < 16$
24. $7 > 0$
25. $7 > -16$
26. $-9 < -8$
27. $-7 > -16$
28. $-4 = -4$
29. $0 > -3$
30. $93 > -100$
31. $30 = 30$
32. $-13 < -3$
33. $24 < 25$
34. $13 > 3$
35. $-24 > -25$

4.2 Absolute Value

1. $|-4| = 4$
2. $|15| = 15$
3. $|0| = 0$
4. $|-5| = 5$
5. $|17| = 17$
6. $|27| = 27$
7. $|-27| = 27$
8. $|8| = 8$
9. $|734| = 734$
10. $|-32| = 32$
11. $|27| = 27$
12. $|1| = 1$
13. $|-9| = 9$
14. $|-731| = 731$
15. $|3| = 3$
16. $|-11| = 11$
17. $|1,369| = 1,369$
18. $|-2| = 2$
19. $|-6| = 6$
20. $|75| = 75$
21. $-|-8| = -8$
22. $|23| = 23$
23. $|-1| = 1$
24. $|-6| = 6$
25. $-|4,683| = -4,683$
26. $-|9| = -9$
27. $-|15| = -15$
28. $-|-94| = -94$
29. $|5| = 5$
30. $-|-12| = -12$
31. $-(3) = -3$
32. $-(26) = -26$
33. $-(1) = -1$
34. $(-4) = -4$
35. $(9) = 9$
36. $-(-1) = 1$
37. $-(-17) = 17$
38. $-(-73) = 73$
39. $-(19) = -19$

40.   (23) = 23
41.   (14) = 14
42.   -(-64) = 64
43.   (-23) = -23
44.   (78) = 78
45.   (-304) = -304
46.   -(23) = -23
47.   (-93) = -93
48.   -(42) = -42
49.   -(-23) = 23
50.   (-2) = -2
51.   (-10) = -10
52.   (8) = 8
53. |-7| = -7
54.   -(8) = -8
55.   -(-5) = 5
56.   -|4| = -4
57.   -|-13| = -13
58.   (93) = 93
59. |4| = 4
60.   -|67| = -67
61. |1| = 1
62.   -(5) = -5
63.   -(-99) = 99
64. |-11| = 11
65.   -|-2| = -2
66.   -(77) = -77
67.   -|83| = -83
68.   (14) = 14
69. |55| = 55
70.   (-6) = -6

4.3 *Addition of Integers*

1.
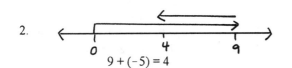
$$4 + 3 = 7$$

2.

$$9 + (-5) = 4$$

3.
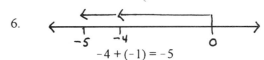
$$5 + (-2) = 3$$

4.
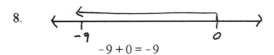
$$-7 + 3 = -4$$

5.
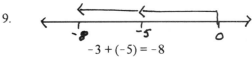
$$-8 + 4 = -4$$

6.
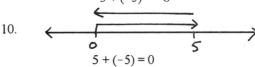
$$-4 + (-1) = -5$$

7.
$$-6 + (-4) = -10$$

8.
$$-9 + 0 = -9$$

9.
$$-3 + (-5) = -8$$

10.
$$5 + (-5) = 0$$

11.   (+5) + (+9) = 14
12.   (-5) + (-9) = -14
13.   (+5) + (-9) = -4
14.   (-5) + (+9) = 4
15.   (-8) + (-6) = -14
16.   9 + (-9) = 0
17.   9 + (-16) = -7
18.   -1 + (-16) = -17
19.   -16 + 12 = -4
20.   9 + (-19) = -10
21.   (-12) + (-18) = -30
22.   25 + 85 = 110
23.   93 + (-102) = -9

21

24. $0 + (-150) = -150$
25. $-117 + 286 = 169$
26. $-51 + 36 = -15$
27. $-78 + 49 + 78 = 49$
28. $-2 + 5 + (-3) = 0$
29. $97 + 3 + (-200) = -100$
30. $(-92) + 16 + (-35) = -111$
31. $63 + 36 + (-59) = 40$
32. $(-23) + (-37) + 43 = -17$
33. $93 + (-45) + (-15) + 27 = 60$
34. $(-87) + (-4) + 9 + (-6) = -88$
35. $427 + (-384) + 749 + (-257) = 535$
36. $-8 + 3 = -5$
37. $0 + (-8) = -8$
38. $8 + (-7) = 1$
39. $20 + [-14) + (-10)]$
       $= 20 + [-24] = -4$
40. $(-6) + (-9) + (-5) = -20$
41. $-4 + 5 + (-2) = -1$
42. $(-11) + (-9) + 7 = -13$
43. $[8 + (-8)] + (-7) = 0 + (-7) = -7$
44. $18 + (-15) + 17 = 20$
45. $35 + (-50) + 10 + (-5) = -10$
46. $-6 + (-11) = -17$
47. $15 + (-32) = -17$
48. $(-8) + (12) = 4$
49. $(-11) + 11 = 0$
50. $0 + (-21) = -21$
51. $(-6) + (-8) = -14$
52. $(-12) + (-3) = -15$
53. $(-18) + 16 = -2$
54. $4 + (-9) = -5$
55. $7 - (-4) = 7 + 4 = 11$
56. $(-10) - (-6) = -10 + 6 = -4$
57. $17 + (-7) = 10$
58. $(-8) + (-9) + 27 = 10$
59. $-2° - 12° = -2 + (-12) = -14°F$
60. $-6° + 56° = 50°F$
61. $-\$26 + (-\$41) + (-\$12)$
       $= -26 + (-41) + (-12) = 79 = \$79$
62. $42 \text{ yards} + 3 \text{ yards} + (-5 \text{ yards})$
       $= 42 + 3 + (-5) = 40 = 40 \text{ yards}$
63. $-24° + 116° = -24 + 116 = 92 = 92°F$
64. $32 \text{ yards} + 8 \text{ yards} + (-10 \text{ yards})$
       $= 32 + 8 + (-10) = 30 = 30 \text{ yards}$

4.4   *Subtraction of Integers*

1. $6 - 4 = 6 + (-4) = 2$
2. $6 - (-4) = 6 + 4 = 10$
3. $-6 - (-4) = -6 + 4 = -2$
4. $-6 - 4 = -6 + (-4) = -10$
5. $7 - (-7) = 7 + 7 = 14$
6. $0 - 2 = 0 + (-2) = -2$
7. $3 - 7 = 3 + (-7) = -4$
8. $40 - (-25) = 40 + 25 = 65$
9. $0 - (-5) = 0 + 5 = 5$
10. $-12 - 7 = -12 + (-7) = -19$
11. $-8 - (-15) = -8 + 15 = 7$
12. $-8 - (-8) = -8 + 8 = 0$
13. $-9 - 14 = -9 + (-14) = -23$
14. $-3 - (-5) = -3 + 5 = 2$
15. $(-6) - 7 = -6 + (-7) = -13$
16. $4 - 7 = 4 + (-7) = -3$
17. $-2 - (-7) = -2 + 7 = 5$
18. $86 - 77 = 86 + (-77) = 9$
19. $-137 - (-142) = -137 + 142 = 5$
20. $(+5) - (-6) = 5 + 6 = 11$
21. $-42 - (-42) = 0$
22. $3 - 10 + 7 - 4 - 5 + 9$
      $= 3 + (-10) + 7 + (-4) + (-5) + 9$
      $= 0$
23. $-15 - 3 = -15 + (-3) = -18$
24. $-9 - (-3) = -9 + 3 = -6$
25. $-10 - 6 + 3 = -10 + (-6) + 3 = -13$
26. $-(-3) - 5 = 3 + (-5) = -2$
27. $6 - 9 = 6 + (-9) = -3$
28. $-15 - 6 = -15 + (-6) = -21$
29. $5 - (-4) - [3 - 4]$
      $= 5 + 4 - [3 + (-4)]$
      $= 5 + 4 - [-1] = 5 + 4 + 1 = 10$
30. $4 - (5 + 2)$
      $= 4 - (7) = 4 + (-7) = -3$
31. $-(-2) - [4 - 6]$
      $= 2 - [4 + (-6)]$
      $= 2 - (-2) = 2 + 2 = 4$
32. $-86 - 77 = -86 + (-77) = -163$
33. $49 - 69 = 49 + (-69) = -20$
34. $24 - (-24) = 24 + 24 = 48$
35. $-2 - 7 = -2 + (-7) = -9$

36. $(5 + 3) - (6 - 2) = 8 - 4 = 4$
37. $-16 - (-16) = -16 + 16 = 0$
38. $-11 - 4 - 15 = -11 + (-4)$
    $+ (-15) = -30$
39. $-217 - 13 = -217 + (-13) = -230$
40. $-36 - (-4) = -36 + 4 = -32$
41. $8 - (-3) = 8 + 3 = 11$
42. $-6 - (-4) = -6 + 4 = -2$
43. $10 - (-9) = 10 + 9 = 19$
44. $-7 - 3 = -7 + (-3) = -10$
45. $-8 - 7 = -8 + (-7) = -15$
46. $-27 - 10 = -27 + (-10) = -37$
47. $7 + (-3) - (-9)$
    $= 7 + (-3) + 9 = 13$
48. $6 - (-3) = 6 + 3 = 9$
49. $-12 + 9 - (-2)$
    $= -12 + 9 + 2 = -1$
50. $-9 - (-3) = -9 + 3 = -6$
51. $28,340 - (-271) = 28,340 + 271$
    $= 28,611$ ft.
52. $1,873 - (-37) = 1,873 + 37 = 1,910$

4.5 Multiplication of Integers

1. $6(4) = 24$
2. $6(-4) = -24$
3. $(-6)(-4) = 24$
4. $(-6)(4) = -24$
5. $(-7)(-3) = 21$
6. $8(-4) = -32$
7. $(9)7 = 63$
8. $(-6)(-7) = 42$
9. $12(-3) = -36$
10. $(-11)(4) = -44$
11. $4(-5)(1) = -20$
12. $4(-5)(-1) = 20$
13. $2(-3)(-1)(4) = 24$
14. $5(5)(-3) = -75$
15. $6(-9) = -54$
16. $(-1)(24)(0) = 0$
17. $7(-7)(2) = -98$
18. $9(-9) = -81$
19. $(-4)(-3)(-2)(4) = -96$
20. $11(-11) = -121$
21. $5(-5)(5) = -125$
22. $4(2)6 = 48$
23. $-21 \cdot 4 = -84$
24. $-3(-33) = 99$
25. $[-16 \cdot 9](-2) = -144(-2) = 288$
26. $(-24)(0) = 0$
27. $(25)(11) = 275$
28. $(-4)(-9)(-6) = -216$
29. $0(05)(-4) = 0$
30. $[-39][-29] = 1,131$
31. $(-5)(45) = -225$
32. $(-25)(12)(1) = -300$
33. $2(-3)(-33) = 198$
34. $[-102][12] = -1,224$
35. $(6)(7)(-4) = -168$
36. $(0)(8)(96) = 0$
37. $(-3)(-107)(0) = 0$
38. $(64)(4 \cdot 2) = (64)8 = 512$
39. $19 \cdot 10 \cdot 8 = 1,520$
40. $(1)(-1)(1)(-1) = 1$
41. $(8)(-3)(-4) = 96$
42. $(5)(5)(4) = 100$
43. $(100)(1)(0) = 0$
44. $(13)(2)(-1) = -26$
45. $(-9)(-8) = 72$
46. $7(-3) = -21$
47. $12(-3) = -36$
48. $(-11)(10) = -110$
49. $[3 + (-5)]7 = (-2)7 = -14$
50. $[-8 - (-2)]27 = [-8 + 2]27$
    $= (-6)27 = -162$
51. $(297)7 = 2,079 = \$2,079$ lost.
52. $(-8)6 = -48 =$ He lost 48 yards.

4.6 Division of Integers

1. $12 \div 3 = 4$
2. $-12 \div 3 = -4$
3. $12 \div (-3) = -4$
4. $-12 \div (-3) = 4$
5. $\dfrac{-27}{3} = -9$

23

6. $\dfrac{72}{-9} = -8$

7. $\dfrac{-42}{-6} = 7$

8. $5\overline{)45}$ with quotient 9

9. $-4\overline{)-36}$ with quotient 9

10. $8\overline{)-32}$ with quotient -4

11. $\dfrac{49}{-7} = -7$

12. $-9 \div (-63) = 7$

13. $11\overline{)121}$ with quotient 11

14. $90 \div (-5) = -18$

15. $\dfrac{-56}{-7} = 8$

16. $9 \div (-1) = -9$

17. $\dfrac{-63}{0} = undefined$

18. $\dfrac{0}{-1} = 0$

19. $144 \div 12 = 12$

20. $\dfrac{-28}{4} = -7$

21. $-17\overline{)187}$ with quotient -11

22. $\dfrac{-625}{-25} = 25$

23. $288 \div (-16) = -18$

24. $\dfrac{225}{-15} = -15$

25. $3\overline{)147}$ with quotient 49

26. $-19\overline{)-418}$ with quotient 22

27. $\dfrac{710}{-710} = -1$

28. $-13\overline{)169}$ with quotient -13

29. $-400 \div 8 = -50$

30. $\dfrac{329}{0} = undefined$

31. $-54 \div (-9) = 6$

32. $36 \div (-3) = -12$

33. $48 \div 6 = 8$

34. $55 \div (-11) = -5$

35. $-60 \div (-12) = 5$

36. $72 \div (-9) = -8$

37. $-99 \div (-11) = 9$

38. $-143 \div (-13) = 11$

39. $-204 \div (-17) = 12$

40. $-361 \div (-19) = 19$

4.7 Order of Operations and Averages Using Integers

1. $3 + 7[9 - (3 - 5)] = 3 + 7[9 - (-2)]$
   $= 3 + 7[9 + 2] = 3 + 7(11)$
   $= 3 + 27 = 80$

2. $5[-6 + 9(-3 + 2)] - 8$
   $= 5[-6 + 9(-1)] - 8$
   $= 5[-6 - 9] - 8 = 5[-1] - 8$
   $= -75 - 8 = -83$

3. $4[9 - (7 + 4)] = 4[9 - 11]$
   $= 4[-2] = -8$

4. $6 - 4[15 - 4(5 - 7)]$
   $= 6 - 4[15 - 4(-2)] = 6 - 4[15 + 8]$
   $= 6 - 4(23) = 6 - 92 = -86$

5. $7 + [(9 - 7) + (-2 - 3)]$
   $= 7 + [2 + (-5)] = 7 + (-3) = 4$

6. $9[(3 - 5)(-4)] = 9[(-2)(-4)]$
   $= 9[8] = 72$

7. $3\{4[-30 - 5(-9 + 3)]\}$
   $= 3\{4[-30 - 5(-6)]\}$
   $= 3\{4[-30 + 30]\} = 3\{4[0]\} = 0$

8. $7 + 6[3(-5) + 4] = 7 + 6[-15 + 4]$
   $= 7 + 6[-11] = 7 - 66 = -59$

9. $[6 - 3(5 - 9)]7 = [6 - 3(-4)]7$
   $= [6 + 12]7 = (18)7 = 126$

10. $9 - 2[(7 - 10) - 2] = 9 - 2[-3 - 2]$
    $= 9 - 2(-5) = 9 + 10 = 19$

11. $-10[(-2^2 + 7) + (-4 + (-10))]$
    $= -10[(-4 + 7) + (-14)]$
    $= -10[3 + (-14)] = -10[-11] = 110$

12. $4[(3 + 8)(-5 + 5)] = 4[(11)(0)]$
    $= 4[0] = 0$

13. $-2 - [(8 - 1)(5)] = -2 - [(7)(5)]$
    $= -2 - (35) = -2 + (035) = -37$

14. $3 - [6 - 3(5 - 9)] = 3 - [6 - 3(-4)]$
    $= 3 - [6 + 12] = 3 - 18 = -15$

15. $17(4) = -68$

16. $(-5(-2) = 10$

17. $(-48) \div 4 = -12$

18. $(-24) \div (-6) = 4$

19. $(-5) + (-13) = -18$

20. $-5 - 13 = -5 + (-13) = -18$

21. $-5 - (-13) = -5 + 13 = 8$

22. $-8 + 2 = -6$

23. $8 + (-2) = 6$

24. $8 - 2 = 6$

25. $(-2)^2 = (-2)(-2) = 4$

26. $(-9)^2 = (-9)(-9) = 81$

27. $(-1)^3 = (-1)(-1)(-1) = -1$

28. $(-4)^2 = (-4)(-4) = 16$

29. $(-7)^3 = (-7)(-7)(-7) = -343$

30. $(-8)^3 = (-8)(-8)(-8) = -512$

31. $(-3)^4 = (-3)(-3)(-3)(-3) = 81$

32. $(-4)^2(3)^2 = (-4)(-4)(3)(3) = 144$

33. $(-2)^2(-3)^2 = (-2)(-2)(-3)(-3) = 36$

34. $(6)^2(-2) = (6)(6)(-2)(-2)(-2) = -288$

35. $(-4)(-2)(-1)(2)(-2) = 32$

36. $\dfrac{-36}{0} = undefined$

37. $-14 - 3 + 7 + 41 = 29$

38. $6 - 13 + 5 - 7 - 9 - 10 = -28$

39. $2(5 - 6)^2 + 7 - 3 = 2(-1)^2 + 7 - 3$
    $= 2(1) + 7 - 3 = 2 + 7 - 3 = 6$

40. $[3 + 7(2^2 - 3)] - 6 = [3 + 7(4 - 3)] - 6$
    $= [3 + 7(1)] - 6 = [3 + 7] - 6$
    $= 10 - 6 = 4$

41. $[3 + 7(2 - 3)^2]6 = [3 + 7(-1)^2]6$
    $= [3 + 7(1)]6 = (10)6 = 60$

42. $-6[3 + 7(2 - 3)] = -6[3 + 7(-1)]$
    $= -6[3 + (-7)] = -6[-4] = 24$

43. $\dfrac{4(5+3)+3}{2(3)+1} = \dfrac{4(8)+3}{6+1} = \dfrac{32+3}{7} =$

    $\dfrac{32+3}{7} = \dfrac{35}{7} = 5$

44. $\dfrac{5(-2)-3(4)}{-2[3-(-2)]-1} = \dfrac{-10-12}{-2[3+2]-1}$

    $\dfrac{-22}{-2[5]-1} = \dfrac{-22}{-10-1} = \dfrac{-22}{-11} = 2$

45. $(-3)(-15) = 45$

46. $13(-4)^2 = 13(16) = 208$

47. $(-54) \div (-9) = 6$

25

48. $\dfrac{72}{-8} = -9$

49. $-3 + 13 = 10$

50. $-3 - (-13) = -3 + 13 = 10$

51. $-12 + (-10) = -22$

52. $-12 - 10 = -12 + (-10) = -22$

53. $4 + (-11) = -7$

54. $4 - 11 = 4 + (-11) = -7$

55. $5(-3)(4)(-1)(-2) = -120$

56. $\dfrac{0}{-36} = 0$

57. $5 - 13 - 19 + 7 - (8)$
$= 5 + (-13) + (-19) + 7 - 64$
$= 5 + (-13) + (-19) + 7 + (-64) = -84$

58. $9 - (3)^2 - 19 + 7 - 8$
$= 9 - 9 + (-19) + 7 + (-8)$
$= 9 + (-9) + (-19) + 7 + (-8) = -20$

59. $12(-2^2) + [3 + (-4)] = 12(-4) + [-7]$
$= -48 + (-7) = -55$

60. $-5[2 + 9(3 - 6)] = -5[2 + 9(-3)]$
$= -5[2 + (-27)] = -5(-25) = 125$

61. $[2 + 9(3 - 6)] - 5 = [2 + 9(-3)] - 5$
$= [2 + (-27) - 5 = -25 + (-5) = -30$

62. $[2 + 9(3 - 6)]5 = [2 + 9(-3)]5$
$= [2 + (-27)]5 = (-25)5 = -125$

63. $\dfrac{8(4+2)-6}{-8-5(-3)} = \dfrac{8(6)-6}{-8+15} =$

$\dfrac{48-6}{7} = \dfrac{42}{7} = 6$

64. $\dfrac{7(5)-6(-4)-(-1)}{5[-2-4]} = \dfrac{35+24+1}{5(-6)} =$

$\dfrac{60}{-30} = -2$

65. $9 - 3^2 = 9 - 9 = 0$

66. $9 + (-3)^2 = 9 + 9 = 18$

67. $4^2 - 4^2 = 16 - 16 = 0$

68. $3^3 - 14 = 27 - 14 = 13$

69. $-3^3 - 14 = -27 - 14 = -41$

70. $9^1 - 5^2 = 9 - 25 = 9 + (-25) = -16$

71. $18^0 - 6^2 = 1 - 36 = -35$

72. $4(3^2 - 3^2) = 4(9 - 9)$
$= 4(0) = 0$

73. $5 + 2^2(3^2 + 4) = 5 + 4(9 + 4)$
$= 5 + 4(13) = 5 + 52 = 57$

74. $3^2 + 8 - 3^2 = 9 + 8 - 9 = 8$

75. $-\$4 + (-\$7) + \$12 + \$9 + (-\$5)$
$= [-4 + (-7) + 12 + 9 + (-5)] \div 5$
$= 5 \div 5 = 1$
The stock gained an average of \$1.

76. $-2° + 15° + (-1°)$
$= [-2 + 15 + (-1)] \div 3$
$= 12 \div 3 = 4$
The average temperature was $4°$F

77. $18° + 8° + (-2°)$
$= [-18 + 8 + (-2)] \div 3$
$= -12 \div 3 = -4$
The average temperature was $-4°$F

78. $-\$7 + \$2 + \$11 + (-\$9) + (-\$12)$
$= [-7 + 2 + 11 + (-9) + (-12)] \div 5$
$= -15 \div 5 = -3$
The stock lost an average of \$3 during the week.

79. $[-20 + (-15) + 10 + (-3) + 19 + (-3)]$
$\div 6 = -12 \div 6 = -2$
Two overbooked seats

80. $[-10 + (-8) + (-5) + (-4) + (-3) + 0 + 2 + 4] \div 8 = -24 \div 8 = -3$
Average score was 3 below par.

4.8 Solving Equations

1. Identity property of equality
2. Symmetric property of equality
3. Transitive property of equality

4. $x + 5 = 8$
$-5 = -5$
$x = 3$

5. $y + 7 = 12$
$-7 = -7$
$y = 5$

6. $x - 98 = 100$
$98 = 98$
$x = 198$

7. $s - 76 = 92$
$76 = 76$
$s = 168$

8. $z + 8 = 6$
$-8 = -8$
$z = -2$

9. $w + 6 = 2$
$-6 = -6$
$w = -4$

10. $y + 5 = -2$
$-5 = -5$
$y = -7$

11. $x + 4 = -6$
$-4 = -4$
$x = -10$

12. $5 = y + 13$
$-13 = -13$
$-8 = y$

13. $8 = p + 12$
$-12 = -12$
$-4 = p$

14. $x - (-2) = -8$
$x + 2 = -8$
$-2 = -2$
$x = -10$

15. $y - (-5) = -3$
$y + 5 = -3$
$-5 = -5$
$y = -8$

16. $0 = y + 34$
$-34 = -34$
$-34 = y$

17. $0 = x + 43$
$-43 = -43$
$-43 = x$

18. $0 = a - 7$
$7 = 7$
$7 = a$

19. $0 = c - 23$
$23 = 23$
$23 = c$

20. $q + 50 = q$
$-q = -q$
$50 = 0$
No solution

21. $z - 70 = z$
$-z = -z$
$-70 = 0$
No solution

22. $a + 37 = a$
$-a = -a$
$37 = 0$
No solution

23. $d - 87 = 3d - 2d$
$d - 87 = d$
$-d = -d$
$-87 = 0$

No solution

24.
$$3 + s = s + 8 - 5$$
$$3 + s = s + 3$$
$$-3 = -3$$
$$s = s$$

Any real number

25.
$$4 + r = r + 48 \div 12$$
$$4 + r = r + 4$$
$$-4 = -4$$
$$r = -4$$
$$r = r$$

Any real number

26.
$$12 - 3 - x = 9 - x$$
$$9 - x = 9 - x$$
$$-9 = -9$$
$$-x = -x$$

Any real number

27.
$$22 - x = 32 - 10 - x$$
$$22 - x = 22 - x$$
$$-22 = -22$$
$$-x = -x$$

any real number

28.
$$19 + a = 6$$
$$-19 = -19$$
$$a = -13$$

29.
$$13 + 12b - 11b = 3$$
$$13 + b = 3$$
$$-13 = -13$$
$$b = -10$$

30.
$$15 + c = -5$$
$$-15 = -15$$
$$c = -20$$

31.
$$9f - 8f - 11 = 33$$
$$f - 11 = 33$$
$$+11 = +11$$
$$f = 44$$

32.
$$z + 2 = 41 - 27$$
$$z + 2 = 14$$
$$-2 = -2$$
$$z = 12$$

33.
$$4q + (-3q) + 8 = 42$$
$$q + 8 = 42$$
$$-8 = -8$$
$$q = 34$$

34.
$$x + 17 = -5 + (-20)$$
$$x + 17 = -25$$
$$-17 = -17$$
$$x = -42$$

35.
$$2b - b - 10 = 6$$
$$b - 10 = 6$$
$$+10 = +10$$
$$b = 16$$

36.
$$x = \textit{number of days per quarter}$$
$$x + 3 = 48$$
$$-3 = -3$$
$$x = 45 \textit{ days per quarter}$$

37.
$$x = \textit{cash going}$$
$$2x = \textit{cash coming}$$
$$x + 15 = 2x$$
$$-x = -x$$
$$15 = x$$

$15 cash in your pocket going to work

REVIEW

28

10.	See Section 4.8	55.	$(-7) + (-9) = -16$
11.	See Section 4.2	56.	$8 + (-6) = 2$
12.	See Section 4.2	57.	$-6 + 16 = 10$
13.	See Section 4.3	58.	$-8 + (-6) = -14$
14.	See Section 4.4	59.	$(-14) + (18) = -32$
15.	See Section 4.5	60.	$-3 + (-5) + 7 + (-8) = -9$
16.	See Section 4.6	61.	$15 + (-15) = 0$
17.	$-3 > -23$	62.	$8 - 6 = 8 + (-6) = 2$
18.	$-3 = -3$	63.	$17 - (-8) = 17 + 8 = 25$
19.	$0 > -3$	64.	$-8 - 6 = -8 + (-6) = -14$
20.	$19 = 19$	65.	$(-3) - 3 = -3 + (-3) = -6$
21.	$-92 < 27$	66.	$8 - (-6) = 8 + 6 = 14$
22.	$-4 < 4$	67.	$14 - 2 = 14 + (-2) = 12$
23.	$-10 < 6$	68.	$-8 - (-6) = -8 + 6 = -2$
24.	$-5 > -10$	69.	$-13 - (-13) = -13 + 13 = 0$
25.	$0 < 15$	70.	$7 - (-3) - [4 - 6]$
26.	$0 > -15$		$= 7 + 3 - [4 + (-6)] = 7 + 3 - [-2]$
27.	$\mid 3 \mid = 3$		$= 7 + 3 + 2 = 12$
28.	$\mid -54 \mid = 54$	71.	$-18 + 15 - [6 - 9]$
29.	$\mid -6 \mid = 6$		$= -18 + 15 - [6 + (-9)]$
30.	$\mid -86 \mid = 86$		$= -18 + 15 - [-3] = -8 + 15 + 3 = 0$
31.	$\mid 342 \mid = 342$	72.	$-7 - 14 - (-2) = -7 + (-14) + 2 = -19$
32.	$\mid -732 \mid = 732$	73.	$13 - 4 - (-3) = 13 + (-4) + 3 = 12$
33.	$\mid 693 \mid = 693$	74.	$-18 - (-18) - 4 = -18 + 18 + (-4) = -4$
34.	$\mid 67 \mid = 67$	75.	$7 - 8 - (2 - 1) = 7 + (-8) - (2 + (-1))$
35.	$\mid 93 \mid = 93$		$= 7 + (-8) - 1 = 7 + 8 + (-8) + (-1)$
36.	$\mid 0 \mid = 0$		$= -2$
37.	7 is the opposite of -7.	76.	$(2 - 8) - 6 - 9$
38.	-63 is the opposite of 63.		$= (2 + (-8)) + (-6) + (-9)$
39.	-14 is the opposite of 14.		$= -6 + (-6) + (-9) = -21$
40.	84 is the opposite of -84.	77.	$8 \cdot 9 = 72$
41.	0 is its own opposite.	78.	$3(0)(-6) = 0$
42.	$-\mid -9 \mid = -(9) = -9$	79.	$(-8)(-9) = 72$
43.	$-(-76) = 76$	80.	$3(-1)(-6) = 18$
44.	$-\mid 23 \mid = -(23) = -23$	81.	$8(-9) = -72$
45.	$-(53) = -53$	82.	$-3(-6)(-9) = -162$
46.	$-\mid 6 \mid = -(6) = -6$	83.	$(-8)9 = -72$
47.	$-(-6) = 6$	84.	$2(7)(9) = 126$
48.	$-\mid -17 \mid = -(17) = -17$	85.	$(-3)(-6)(-1)(10) = -180$
49.	$-(27) = -27$	86.	$[-7 \cdot 11](-2) = (-72)(-2) = 154$
50.	$(19) = 19$	87.	$64(-8)(0) = 0$
51.	$\mid -3 \mid = 3$	88.	$8(1)(-4)(-3) = 96$
52.	$8 + 6 = 14$	89.	$-(8)(5)(2) = -80$
53.	$13 + (-6) = 7$	90.	$5 \cdot (6)(-3) = -90$
54.	$-8 + 6 = -2$	91.	$(-4)(3 \cdot 9) = -108$

92. $35 \div 7 = 5$

93. $-72 \div 9 = -8$

94. $\dfrac{-35}{-7} = 5$

95. $99 \div (-11) = -9$

96. $7 \overline{\smash{)}-35}^{\,-5}$

97. $-323 \div (-17) = 19$

98. $35 \div (-7) = -5$

99. $-100 \div 10 = -10$

100. $\dfrac{-6}{0} = undefined$

101. $102 \div (-3) = -34$

102. $-434 \div 7 = -62$

103. $115 \div (-5) = -23$

104. $-88 \div (-8) = 11$

105. $-98 \div 7 = -14$

106. $-7^2 = (-7)(-7) = -49$

107. $(-7)^2 = (-7)(-7) = 49$

108. $(8-6)^2 = (2)^2 = (2)(2) = 4$

109. $8 - 6^2 = 8 - (6)(6) = 8 - 36$
$= 8 + (-36) = -28$

110. $2 + 6[4^2 - (2-4)]$
$= 2 + 6[16 - (-2)]$
$= 2 + 6[16 + 2] = 2 + 6[18]$
$= 2 + 108 = 110$

111. $[8 + 2(3-6)]6 = [8 + 2(-3)]6$
$= [8 + (-6)]6 = [2]6 = 12$

112. $4 + [(8 - 2^2) + (-6 - 8)]$
$= 4 + [(8-4) + (-14)]$
$= 4 + [4 + (-14)] = 4 + (-10) = -6$

113. $\dfrac{4(3+2)+4}{-4-2} = \dfrac{4(5)+4}{-4+(-2)} =$

$\dfrac{20+4}{-6} + \dfrac{24}{-6} = -4$

114. $(5^2 - 25)0$
$= (25 - 25)0$
$= (0)0 = 0$

115. $27 + 3^2 - 6(-5)$
$= 27 + 9 + 30 = 66$

116. $(17 - 3)(3^2 - 9)$
$= (14)(9 - 9) = 14(0) = 0$

117. $3(8 - 4) - 7(2^2 + 4)$
$= 3(4) - 7(4 + 4) = 12 - 7(8)$
$= 12 - 56 = 12 + (-56) = -44$

118. $4(3 - 2^2) - 4(3 - 2)^2$
$= 4(3 - 4) - 4(1)^2 = 4(-1) - 4(1)$
$= -4 - 4 = -4 + (-4) = -8$

119. $3 + 4[2^2 - (6 - 8)]$
$= 3 + 4[4 - (6 + (-8))]$
$= 3 + 4[4 - (-2)] = 3 + 4[4 + 2]$
$= 3 + 4[6] = 3 + 24 = 27$

120. $\dfrac{7(4+2)+3}{-3-2} = \dfrac{7(6)+3}{-3+(-2)} = \dfrac{42+3}{-5} =$

$\dfrac{45}{-5} = -9$

121. $\{-29, -35, 7\}$

$\dfrac{-29+(-35)+7}{3} = \dfrac{-57}{3} = -19$

122. $\{5, -19, 0, -12, 6\}$

$\dfrac{5+(-19)+0+(-12)+6}{5} = \dfrac{-20}{5} = -4$

123. $\{9, -15, 8, 2\}$

$\dfrac{9+(-15)+8+2}{4} = \dfrac{4}{4} = 1$

124. $\{6,-7,9,0\}$

$$\frac{6+(-7)+9+0}{4} = \frac{8}{4} = 2$$

125. $\{19,-9,2\}$

$$\frac{19+(-9)+2}{3} = \frac{12}{3} = 4$$

126.
$$\begin{aligned} x+6 &= 10 \\ -6 &= -6 \\ x &= 4 \end{aligned}$$

127.
$$\begin{aligned} 7y+(-6y)-4 &= 4 \\ y-4 &= 4 \\ 4 &= 4 \\ y &= 8 \end{aligned}$$

128.
$$\begin{aligned} -4a+3+5a &= 6 \\ a+3 &= 6 \\ -3 &= -3 \\ a &= 3 \end{aligned}$$

129.
$$\begin{aligned} 10q &= 10q+3 \\ -10q &= -10q \\ 0 &= 3 \end{aligned}$$

No solution

130.
$$\begin{aligned} 7+t &= 3+t+4 \\ 7+t &= 7+t \\ -7 &= -7 \\ t &= t \end{aligned}$$

Any real number

PRACTICE TEST

1. 0 is greater than -13.

2. $|-6| = 6 \quad |4| = 4$
   6 is greater than 4; $|-6| > |4|$

3. $-|-7| = -(7) = -7$
   $-(-7) = 7$
   -7 is smaller than 7; $-|-7| < -(-7)$

4. $-(-47) = 47$
   $-(47) = -47$
   47 is greater than -47; $-(-47) > -(47)$

5. $(-7)(-8) = 56$

6. $-4 + 17 + (-23) = -10$

7. $15 + (-15) = 0$

8. $-15 + (-15) = -30$

9. $-72 \div 9 = -8$

10. $[6 \cdot 5](-4) = [30](-4) = -120$

11. $-23 \div 0 = $ undefined

12. $6 - 12 = 6 + (-12) = -6$

13. $-6 - 12 = -6 + (-12) = -18$

14. $(-1)(7)(-6)(-2)(-3) = 252$

15. $22 - (-22) = 22 + 22 = 44$

16. $\{15, 0, -12\}$

$$\frac{15+0+(-12)}{3} = \frac{3}{3} = 1$$

17. $-6^2 = -(6)(6) = -36$

18. $(-6)^2 = (-6)(-6) = 36$

19. $6[-5-(-5)](-34) = 6[-5+5](-34)$
    $= 6[0](-34) = 0$

20. $6 + 2[3^2 - (3-7)] = 6 + 2[9 - (-4)]$
    $= 6 + 2[9+4] = 6 + 2[13]$
    $= 6 + 26 = 32$

31

21. $5 + [7^2 - 5(8)]3 = 5 + [49 - 40]3$
       $= 5 + [9]3 = 5 + 27 = 32$

22. $5 + 7^2 - 5(8)] - 3 = 5 + [49 - 40] - 3$
       $= 5 + 9 - 3 = 11$

23. $\dfrac{7(9-5)}{-2^2} = \dfrac{7(4)}{-4} = \dfrac{28}{-4} = -7$

24. $6 + [(3 - 17) + (-2 - 5)]$
       $= 6 + [(3 + (-17)) + (-2 + (-5))]$
       $= 6 + [-14 + (-7)] = 6 + [-21] = -15$

25. $\begin{aligned} x + 9 &= 17 \\ -9 &= -9 \\ x &= 8 \end{aligned}$

26. $\begin{aligned} y - 9 &= 9 \\ 9 &= 9 \\ y &= 18 \end{aligned}$

27. $\begin{aligned} 2a &= 2a + 7 \\ -2a &= -2a \\ 0 &= 7 \end{aligned}$
       No solution

28. $\begin{aligned} 6t &= 9 + 5t \\ -5t &= -5t \\ t &= 9 \end{aligned}$

29. $\begin{aligned} c - 23 &= -13 \\ 23 &= 23 \\ c &= 10 \end{aligned}$

30. $\begin{aligned} x - 8 &= 2x \\ -x &= -x \\ -8 &= x \end{aligned}$

CHAPTER 5: FRACTIONS AND MIXED NUMBERS

5.1 The Set of Rational Numbers as Fractions

1. 3
2. 5
3. 1
4. 3
5. 8
6. 5
7. 3
8. 1
9. 13
10. 9
11. 3
12. 7
13. 2
14. 8
15. 0
16. -2
17. 20
18. 13
19. 5
20. -5
21. Proper fraction

22. $\dfrac{20}{6} = 3\dfrac{2}{6} = 3\dfrac{1}{3}$

23. $\dfrac{6}{1} = 6$

24. $\dfrac{22}{9} = 2\dfrac{4}{9}$

25. Proper fraction

26. $\dfrac{8}{5} = 1\dfrac{3}{5}$

27. Proper fraction

28. $\dfrac{55}{55} = 1$

29. $\dfrac{57}{7} = 8\dfrac{1}{7}$

30. $\dfrac{40}{19} = 2\dfrac{2}{19}$

31. Negative
32. Negative
33. Positive
34. Positive
35. Negative
36. Negative
37. Positive
38. Negative
39. Negative
40. 0

41. $\dfrac{17}{5} = 3\dfrac{2}{5}$

42. $\dfrac{28}{7} = 4$

43. $\dfrac{11}{3} = 3\dfrac{2}{3}$

44. $\dfrac{19}{2} = 9\dfrac{1}{2}$

45. $\dfrac{76}{7} = 10\dfrac{6}{7}$

46. $\dfrac{14}{3} = 4\dfrac{2}{3}$

47. $\dfrac{-10}{3} = -3\dfrac{1}{3}$

48. $\dfrac{20}{8} = 2\dfrac{4}{8} = 2\dfrac{1}{2}$

49. $\dfrac{32}{-3} = -10\dfrac{2}{3}$

50. $\dfrac{82}{9} = 9\dfrac{1}{9}$

51. $\dfrac{106}{3} = 35\dfrac{1}{3}$

52. $\dfrac{-67}{6} = -11\dfrac{1}{6}$

53. $7\dfrac{3}{4} = \dfrac{31}{4}$

54. $9\dfrac{7}{8} = \dfrac{79}{8}$

55. $7\dfrac{2}{9} = \dfrac{65}{9}$

56. $8\dfrac{1}{2} = \dfrac{17}{2}$

57. $4\dfrac{2}{3} = \dfrac{14}{3}$

58. $3\dfrac{9}{13} = \dfrac{48}{13}$

59. $3\dfrac{4}{5} = \dfrac{19}{5}$

60. $10\dfrac{7}{8} = \dfrac{87}{8}$

5.2 Using the Identity Property of Multiplication with Fractions

1. $\dfrac{1}{3} = \dfrac{1}{3} \cdot \dfrac{2}{2} = \dfrac{2}{6}$

2. $\dfrac{1}{2} = \dfrac{1}{2} \cdot \dfrac{3}{3} = \dfrac{3}{6}$

3. $\dfrac{2}{12} = \dfrac{2}{12} \div \dfrac{2}{2} = \dfrac{1}{6}$

4. $\dfrac{8}{24} = \dfrac{8}{24} \div \dfrac{4}{4} = \dfrac{2}{6}$

5. $\dfrac{1}{5} = \dfrac{1}{5} \cdot \dfrac{2}{2} = \dfrac{2}{10}$

6. $\dfrac{1}{2} = \dfrac{1}{2} \cdot \dfrac{5}{5} = \dfrac{5}{10}$

7. $\dfrac{6}{30} = \dfrac{6}{30} \div \dfrac{3}{3} = \dfrac{2}{10}$

8. $\dfrac{18}{20} = \dfrac{18}{20} \div \dfrac{2}{2} = \dfrac{9}{10}$

9. $\dfrac{1}{4} = \dfrac{1}{4} \cdot \dfrac{3x}{3x} = \dfrac{3x}{12x}$

10. $\dfrac{2}{2x} = \dfrac{2}{2x} \cdot \dfrac{6}{6} = \dfrac{12}{12x}$

11. $\dfrac{8}{24x} = \dfrac{8}{24x} \div \dfrac{2}{2} = \dfrac{4}{12x}$

12. $\dfrac{7}{12} = \dfrac{7}{12} \cdot \dfrac{x}{x} = \dfrac{7x}{12x}$

13. GCF$(3y, 3y^2)$ 3y   , $3y^2$

    GCF$(3y, 3y^2) = 3y$

14. GCF$(7x, 14x)$ 7x   ,   14x

    GCF$(7x, 14x) = 7x$

15. GCF$(2ab, 2a^2)$ 2ab   , $2a^2$

    GCF$(2ab, 2a^2) = 2a$

34

16. GCF (x^2y, y^2) x^2y , y^2

$x \cdot x \cdot y$ $y \cdot y$

GCF $(x^2y, y^2) = y$

17. $\dfrac{6}{12} = \dfrac{2 \cdot 3}{2 \cdot 2 \cdot 3} = \dfrac{1}{2}$

18. $\dfrac{12}{6} = \dfrac{2 \cdot 2 \cdot 3}{2 \cdot 3} = \dfrac{2}{1} = 2$

19. $\dfrac{10}{15} = \dfrac{2 \cdot 5}{3 \cdot 5} = \dfrac{2}{3}$

20. $\dfrac{7}{8} = \dfrac{7}{8}$

21. $\dfrac{20}{25} = \dfrac{2 \cdot 2 \cdot 5}{5 \cdot 5} = \dfrac{4}{5}$

22. $\dfrac{27}{72} = \dfrac{3 \cdot 3 \cdot 3}{2 \cdot 2 \cdot 2 \cdot 3 \cdot 3} = \dfrac{3}{8}$

23. $\dfrac{11}{99} = \dfrac{11}{3 \cdot 3 \cdot 11} = \dfrac{1}{9}$

24. $\dfrac{21}{49} = \dfrac{3 \cdot 7}{7 \cdot 7} = \dfrac{3}{7}$

25. $\dfrac{28}{56} = \dfrac{2 \cdot 2 \cdot 7}{2 \cdot 2 \cdot 2 \cdot 7} = \dfrac{1}{2}$

26. $\dfrac{19}{38} = \dfrac{19}{2 \cdot 9} = \dfrac{1}{2}$

27. $\dfrac{13}{52} = \dfrac{13}{2 \cdot 2 \cdot 13} = \dfrac{1}{4}$

28. $\dfrac{45}{54} = \dfrac{5 \cdot 3 \cdot 3}{2 \cdot 3 \cdot 3 \cdot 3} = \dfrac{5}{6}$

29. $\dfrac{36}{25} = \dfrac{2 \cdot 2 \cdot 3 \cdot 3}{5 \cdot 5} = \dfrac{36}{25}$

30. $\dfrac{24a^2b}{12ab^2} =$

$\dfrac{2 \cdot 2 \cdot 2 \cdot 3 \cdot a \cdot a \cdot b}{2 \cdot 2 \cdot 3 \cdot a \cdot b \cdot b} = \dfrac{2a}{b}$

31. $\dfrac{x^2}{xy} = \dfrac{x \cdot x}{x \cdot y} = \dfrac{x}{y}$

32. $\dfrac{45x}{40x^2} = \dfrac{3 \cdot 3 \cdot 5 \cdot x}{2 \cdot 2 \cdot 2 \cdot 5 \cdot x \cdot x} = \dfrac{9}{8x}$

5.3 Multiplication of Fractions and Mixed Numbers

1. $\dfrac{2}{3} \cdot \dfrac{4}{5} = \dfrac{8}{15}$

2. $\dfrac{4}{5} \cdot \dfrac{1}{4} = \dfrac{2 \cdot 2 \cdot 1}{5 \cdot 2 \cdot 2} = \dfrac{1}{5}$

3. $\dfrac{2}{5} \cdot \dfrac{3}{7} = \dfrac{6}{35}$

4. $\dfrac{4}{9} \cdot \dfrac{9}{4} = \dfrac{1}{1} = 1$

5. $5 \cdot -\dfrac{1}{5} = \dfrac{5}{1} \cdot -\dfrac{1}{5} = -\dfrac{1}{1} = -1$

6. $7 \cdot \dfrac{3}{7} = \dfrac{7}{1} \cdot \dfrac{3}{7} = \dfrac{3}{1} = 3$

7. $-3 \cdot \dfrac{5}{6} = \dfrac{-3}{1} \cdot \dfrac{5}{6} =$

$\dfrac{-3 \cdot 5}{1 \cdot 2 \cdot 3} = -\dfrac{5}{2}$

8. $(\dfrac{2}{5})(-\dfrac{2}{3})(\dfrac{4}{5}) = \dfrac{2 \cdot 2 \cdot 4}{5 \cdot 3 \cdot 5} = -\dfrac{16}{75}$

35

9. $\left(-\dfrac{1}{8}\right)\left(-\dfrac{2}{3}\right)(-8) =$

$$\dfrac{2 \cdot 2 \cdot 2 \cdot 2}{2 \cdot 2 \cdot 2 \cdot 3} = \dfrac{-2}{3}$$

10. $\dfrac{9}{20} \cdot \dfrac{4}{3} = \dfrac{3 \cdot 3 \cdot 2 \cdot 2}{2 \cdot 2 \cdot 4 \cdot 3} = \dfrac{3}{5}$

11. $\dfrac{-27}{40} \cdot \dfrac{8}{3} =$

$$\dfrac{3 \cdot 3 \cdot 3 \cdot 2 \cdot 2 \cdot 2}{2 \cdot 2 \cdot 2 \cdot 5 \cdot 3} = \dfrac{-9}{5}$$

12. $\left(-\dfrac{3}{5}\right)\left(-\dfrac{3}{4}\right) = \dfrac{9}{20}$

13. $\left(-\dfrac{2}{3}\right)\left(\dfrac{1}{5}\right) = \dfrac{2}{15}$

14. $\left(1\dfrac{5}{9}\right)\left(-\dfrac{6}{7}\right) = \left(\dfrac{14}{9}\right)\left(-\dfrac{6}{7}\right) =$

$$-\dfrac{2 \cdot 7 \cdot 2 \cdot 3}{3 \cdot 3 \cdot 7} = -\dfrac{4}{3}$$

15. $\left(-3\dfrac{1}{2}\right)\left(-1\dfrac{1}{7}\right) = \left(-\dfrac{7}{2}\right)\left(-\dfrac{8}{7}\right) =$

$$\dfrac{7 \cdot 2 \cdot 2 \cdot 2}{2 \cdot 7} = 4$$

16. $\left(-\dfrac{9}{11}\right)\left(-7\dfrac{1}{3}\right) = \left(-\dfrac{9}{11}\right)\left(-\dfrac{22}{3}\right) =$

$$\dfrac{3 \cdot 3 \cdot 2 \cdot 11}{11 \cdot 3} = 6$$

17. $\left(1\dfrac{7}{9}\right)\left(\dfrac{27}{32}\right) = \dfrac{16}{9} \cdot \dfrac{27}{32} =$

$$\dfrac{2 \cdot 2 \cdot 2 \cdot 2 \cdot 3 \cdot 3 \cdot 3}{3 \cdot 3 \cdot 2 \cdot 2 \cdot 2 \cdot 2 \cdot 2} = \dfrac{3}{2}$$

18. $\left(-\dfrac{42}{7}\right)\left(-\dfrac{15}{36}\right)\left(-\dfrac{6}{7}\right) =$

$$\dfrac{2 \cdot 3 \cdot 7 \cdot 3 \cdot 5 \cdot 2 \cdot 3}{7 \cdot 2 \cdot 2 \cdot 3 \cdot 3 \cdot 7} =$$

$$-\dfrac{15}{7} \text{ or } -2\dfrac{1}{7}$$

19. $\left(3\dfrac{3}{5}\right)\left(-\dfrac{5}{36}\right) = \left(\dfrac{18}{5}\right)\left(-\dfrac{5}{36}\right) =$

$$-\dfrac{2 \cdot 3 \cdot 3 \cdot 5}{5 \cdot 2 \cdot 2 \cdot 3 \cdot 3} = -\dfrac{1}{2}$$

20. $\left(-\dfrac{7}{8}\right)\left(\dfrac{1}{14}\right)16 =$

$$-\dfrac{7 \cdot 1 \cdot 2 \cdot 2 \cdot 2 \cdot 2}{2 \cdot 2 \cdot 2 \cdot 2 \cdot 7} = -\dfrac{1}{1} = 1$$

21. $\left(3\dfrac{3}{7}\right)\left(-2\dfrac{5}{8}\right)\left(-\dfrac{1}{9}\right) =$

$$\left(\dfrac{24}{7}\right)\left(-\dfrac{21}{8}\right)\left(-\dfrac{1}{9}\right) =$$

$$\dfrac{2 \cdot 2 \cdot 2 \cdot 3 \cdot 3 \cdot 7}{7 \cdot 2 \cdot 2 \cdot 2 \cdot 3 \cdot 3} = \dfrac{1}{1} = 1$$

22. $\left(-\dfrac{14}{45}\right)\left(\dfrac{3}{56}\right) =$

$$\dfrac{2 \cdot 7 \cdot 3}{5 \cdot 3 \cdot 3 \cdot 7 \cdot 2 \cdot 2 \cdot 2} = -\dfrac{1}{60}$$

23. $\left(-\dfrac{15}{30}\right)\left(-1\dfrac{3}{5}\right) = \left(-\dfrac{15}{32}\right)\left(-\dfrac{8}{5}\right)$

$$\dfrac{3 \cdot 5 \cdot 2 \cdot 2 \cdot 2}{2 \cdot 2 \cdot 2 \cdot 2 \cdot 2 \cdot 5} = \dfrac{3}{4}$$

36

24. $10 \cdot \dfrac{1}{100} \left(-\dfrac{1}{10}\right) =$

       $\dfrac{10}{1} \cdot \dfrac{1}{100} \cdot \left(-\dfrac{1}{10}\right) =$

       $\dfrac{2 \cdot 5 \cdot 1 \cdot 1}{1 \cdot 2 \cdot 2 \cdot 5 \cdot 5 \cdot 2 \cdot 5} = -\dfrac{1}{100}$

25. $\left(\dfrac{1}{2}\right)^2 = \dfrac{1}{2} \cdot \dfrac{1}{2} = \dfrac{1}{4}$

26. $\left(\dfrac{2}{3}\right)^3 = \dfrac{2}{3} \cdot \dfrac{2}{3} \cdot \dfrac{2}{3} = \dfrac{8}{27}$

27. $\left(-\dfrac{1}{4}\right)^2 = \left(-\dfrac{1}{4}\right)\left(-\dfrac{1}{4}\right) = \dfrac{1}{16}$

28. $\left(-\dfrac{3}{5}\right)^2 = \left(-\dfrac{3}{5}\right)\left(-\dfrac{3}{5}\right) = \dfrac{9}{25}$

29. $\left(-\dfrac{1}{2}\right)^3\left(-\dfrac{2}{3}\right)^2 =$

       $\left(-\dfrac{1}{2}\right)\left(-\dfrac{1}{2}\right)\left(-\dfrac{1}{2}\right)\left(-\dfrac{2}{3}\right)\left(-\dfrac{2}{3}\right) = -\dfrac{1}{18}$

30. $\left(\dfrac{3}{4}\right)^2\left(-\dfrac{2}{3}\right)^2 =$

       $\left(\dfrac{3}{4}\right)\left(\dfrac{3}{4}\right)\left(-\dfrac{2}{3}\right)\left(-\dfrac{2}{3}\right) = \dfrac{1}{4}$

31. $\dfrac{a^2b}{c} \cdot \dfrac{c^3}{ab^2} =$

       $\dfrac{a \cdot a \cdot b \cdot c \cdot c \cdot c}{c \cdot a \cdot b \cdot b} = \dfrac{ac^2}{b}$

32. $\dfrac{xy^2}{z^2} \cdot \dfrac{z}{x^2y} =$

       $\dfrac{x \cdot y \cdot y \cdot z}{z \cdot z \cdot x \cdot x \cdot y} = \dfrac{y}{xz}$

33. $\left(2\dfrac{1}{2}\right)\left(4\dfrac{2}{3}\right) = \dfrac{5}{2} \cdot \dfrac{14}{3} = \dfrac{35}{3} = 11\dfrac{2}{3}$

34. $\left(5\dfrac{1}{4}\right)\left(-1\dfrac{2}{3}\right) = \dfrac{21}{4} \cdot \left(-\dfrac{5}{3}\right) =$

       $-\dfrac{35}{4} = -8\dfrac{3}{4}$

35. $\left(-7\dfrac{1}{2}\right)\left(-2\dfrac{1}{5}\right)(2) =$

       $\left(-\dfrac{15}{2}\right)\left(-\dfrac{11}{5}\right)\left(\dfrac{2}{1}\right) = \dfrac{33}{1} = 33$

36. $\left(-3\dfrac{5}{9}\right)\left(2\dfrac{5}{8}\right) = \left(-\dfrac{32}{9}\right)\left(\dfrac{21}{8}\right) =$

       $-\dfrac{28}{3} = -9\dfrac{1}{3}$

37. $\left(\dfrac{3}{5}\right)^2\left(-1\dfrac{2}{9}\right)^2 =$

       $\dfrac{3}{5} \cdot \dfrac{3}{5} \cdot \left(-\dfrac{11}{9}\right)\left(-\dfrac{11}{9}\right) = \dfrac{121}{225}$

38. $\dfrac{3ab}{2c^2} \cdot \dfrac{8c}{6b^2} =$

       $\dfrac{3 \cdot a \cdot b \cdot 2 \cdot 2 \cdot 2 \cdot c}{2 \cdot c \cdot c \cdot 2 \cdot 3 \cdot b \cdot b} = \dfrac{2a}{bc}$

39. $\dfrac{x}{y} \cdot \dfrac{y}{x} = \dfrac{x \cdot y}{y \cdot x} = 1$

40. $\dfrac{64x^2}{9y^2} \cdot \dfrac{72y}{16x} =$

       $\dfrac{2 \cdot 2 \cdot 2 \cdot 2 \cdot 2 \cdot 2 \cdot x \cdot x \cdot 2 \cdot 2 \cdot 2 \cdot 3 \cdot 3 \cdot y}{3 \cdot 3 \cdot y \cdot y \cdot 2 \cdot 2 \cdot 2 \cdot 2 \cdot x} =$

       $\dfrac{32x}{y}$

41. $\dfrac{63ab}{21c} \cdot \dfrac{7c^2}{35b^2} =$

$\dfrac{3 \cdot 3 \cdot 7 \cdot a \cdot b \cdot 7 \cdot c \cdot c}{3 \cdot 7 \cdot c \cdot 5 \cdot 7 \cdot b \cdot b} = \dfrac{3ac}{5b}$

42. $\dfrac{55m}{19n} \cdot \dfrac{38mn}{11m^2} =$

$\dfrac{5 \cdot 11 \cdot m \cdot 2 \cdot 19 \cdot m \cdot n}{19 \cdot n \cdot 11 \cdot m \cdot m} = 10$

43. $\dfrac{1}{2} \cdot \dfrac{7}{8} = \dfrac{7}{16}$

44. $\dfrac{1}{4} \cdot 44 = \dfrac{1}{\cancel{4}} \cdot \dfrac{\cancel{44}^{11}}{1} = 11$

45. $\dfrac{4}{5} \cdot \dfrac{3}{4} = \dfrac{3}{5}$

46. $\dfrac{3}{5} \cdot \dfrac{1}{3} \cdot 15 = \dfrac{\cancel{3}}{\cancel{5}} \cdot \dfrac{1}{\cancel{3}} \cdot \dfrac{\cancel{15}^3}{1} = 3$

47. $2 \cdot \dfrac{7}{8} = \dfrac{\cancel{2}}{1} \cdot \dfrac{7}{\cancel{8}_4} = \dfrac{7}{4} =$

1¾ cups of flour.

48. $\dfrac{1}{2} \cdot \dfrac{7}{8} = \dfrac{7}{16}$ of a cup or flour.

49. $\dfrac{1}{3} \cdot \dfrac{3}{4} = \dfrac{1}{4}$ of the class received an A.

50. 1 in 9 is $\dfrac{1}{9}$; $\dfrac{1}{9}$ of 1,737 is

$\dfrac{1}{9} \cdot 1,737 = 193$ women can expect to develop cancer.

51. $\dfrac{1}{2} \cdot \dfrac{3}{4} = \dfrac{3}{8}$ ton

52. $\dfrac{7}{8}$ of the snow left so:

$\dfrac{7}{8} \cdot \dfrac{3}{4} = \dfrac{21}{32}$ inches of snow left.

53. $3 \cdot 2\dfrac{1}{2} = \dfrac{3}{1} \cdot \dfrac{5}{2} = \dfrac{15}{2} =$

7½ cup of flour.

54. $\dfrac{3}{4} \cdot 2\dfrac{1}{2} = \dfrac{3}{4} \cdot \dfrac{5}{2} = \dfrac{15}{8} =$

$1\dfrac{7}{8}$ cup of flour.

55. $\dfrac{1}{4} \cdot \dfrac{7}{8} = \dfrac{7}{32}$ of the class received an A

56. 1 in 9 is $\dfrac{1}{9}$. $\dfrac{1}{9} \cdot 3,663 =$

407 men can expect to develop cancer.

57. $\dfrac{1}{10}$ delivered so $\dfrac{9}{10}$ left

$\dfrac{9}{10} \cdot \dfrac{7}{8} = \dfrac{63}{80}$ ton of gravel left in truck

58. $\dfrac{1}{3}$ melted so $\dfrac{2}{3}$ left.

$\dfrac{2}{3} \cdot 2\dfrac{7}{8} = \dfrac{\cancel{2}}{3} \cdot \dfrac{23}{\cancel{8}_4} = \dfrac{23}{12} =$

$1\dfrac{11}{12}$ inch of snow left.

59. $\dfrac{3}{\cancel{22}} \cdot \cancel{15,246}^{693} = 2,079$ fish.

60. 108 in $\dfrac{1}{3}$ so, $108 \cdot 3 = 324$ fish in the whole lake.

38

5.4 Division of Fractions and Mixed Numbers

1. $\dfrac{1}{2} \div \dfrac{3}{5} = \dfrac{1}{2} \cdot \dfrac{5}{3} = \dfrac{5}{6}$

2. $\dfrac{2}{3} \div \dfrac{5}{7} = \dfrac{2}{3} \cdot \dfrac{7}{5} = \dfrac{14}{15}$

3. $\dfrac{1}{4} \div \left(-\dfrac{3}{8}\right) = \dfrac{1}{\cancel{4}} \cdot \left(-\dfrac{\cancel{8}^2}{3}\right) = -\dfrac{2}{3}$

4. $-\dfrac{4}{5} \div \dfrac{10}{2} = -\dfrac{4}{5} \cdot \dfrac{2}{10} =$

   $-\dfrac{8}{50} = -\dfrac{4}{25}$

5. $\dfrac{1}{3} \div (-6) = \dfrac{1}{3} \div \left(-\dfrac{6}{1}\right) =$

   $\dfrac{1}{3} \cdot \left(-\dfrac{1}{6}\right) = -\dfrac{1}{18}$

6. $\dfrac{-4}{5} \div (-4) = \dfrac{-4}{5} \div \left(-\dfrac{4}{1}\right) =$

   $-\dfrac{4}{5} \cdot \left(-\dfrac{1}{4}\right) = \dfrac{1}{5}$

7. $\dfrac{7}{8} \div \left(-\dfrac{7}{8}\right) = \dfrac{7}{8} \cdot -\dfrac{8}{7} = 1$

8. $-\dfrac{25}{36} \div \dfrac{5}{6} = -\dfrac{\cancel{25}^5}{\cancel{36}^6} \cdot \dfrac{\cancel{6}}{\cancel{5}} = -\dfrac{5}{6}$

9. $\dfrac{3}{4} \div (-3) = \dfrac{3}{4} \div \left(-\dfrac{3}{1}\right) =$

   $\dfrac{3}{4} \cdot \left(-\dfrac{1}{3}\right) = -\dfrac{1}{4}$

10. $\dfrac{14}{3} \div \dfrac{16}{9} = \dfrac{\cancel{14}^7}{\cancel{3}} \cdot \dfrac{\cancel{9}^3}{\cancel{16}_8} = \dfrac{21}{8}$ or $2\dfrac{5}{8}$

11. $-1\dfrac{1}{14} \div \dfrac{5}{2} = -\dfrac{15}{14} \div \dfrac{5}{2} =$

    $-\dfrac{\cancel{15}^3}{\cancel{14}_7} \cdot \dfrac{\cancel{2}}{\cancel{5}} = -\dfrac{3}{7}$

12. $-\dfrac{3}{2} \div \left(-\dfrac{3}{2}\right) = -\dfrac{3}{2} \cdot \left(-\dfrac{2}{3}\right) = 1$

13. $\dfrac{19}{45} \div \left(-5\dfrac{7}{10}\right) = \dfrac{19}{45} \div \left(-\dfrac{57}{10}\right) =$

    $\dfrac{\cancel{19}}{\cancel{45}_9} \cdot \left(-\dfrac{\cancel{10}^2}{\cancel{57}_3}\right) = -\dfrac{2}{27}$

14. $\dfrac{17}{13} \div \left(-\dfrac{17}{13}\right) = \dfrac{17}{13} \cdot \left(-\dfrac{13}{17}\right) = -1$

15. $1\dfrac{1}{14} \div \left(-1\dfrac{17}{28}\right) = \dfrac{15}{14} \div \left(-\dfrac{45}{28}\right) =$

    $\dfrac{\cancel{15}}{\cancel{14}} \cdot -\dfrac{\cancel{28}^2}{\cancel{45}_3} = -\dfrac{2}{3}$

16. $3\dfrac{3}{7} \div \left(-\dfrac{8}{21}\right) = \dfrac{\cancel{24}^3}{\cancel{7}} \cdot \left(-\dfrac{\cancel{21}^3}{\cancel{8}}\right) = -9$

17. $\dfrac{13}{9} \div \dfrac{13}{9} = \dfrac{13}{9} \cdot \dfrac{9}{13} = 1$

18. $\dfrac{17}{20} \div 13\dfrac{3}{5} = \dfrac{17}{20} \div \dfrac{68}{5} =$

    $\dfrac{\cancel{17}}{\cancel{20}_4} \cdot \dfrac{\cancel{5}}{\cancel{68}_4} = \dfrac{1}{16}$

19. $-\dfrac{3}{14} \div \left(-\dfrac{6}{7}\right) = -\dfrac{\cancel{3}^1}{\cancel{14}_2} \cdot -\dfrac{\cancel{7}^1}{\cancel{6}_2} = \dfrac{1}{4}$

20. $\quad \dfrac{1}{4} \div \dfrac{1}{9} = \dfrac{1}{4} \cdot \dfrac{9}{1} = \dfrac{9}{4}$ or $2\dfrac{1}{4}$

21. $\quad \dfrac{x^2y}{z} \div \dfrac{z^2}{xy} = \dfrac{x^2y}{z} \cdot \dfrac{xy}{z^2} =$

$\quad \dfrac{x \cdot x \cdot y \cdot x \cdot y}{z \cdot z \cdot z} = \dfrac{x^3y^2}{z^3}$

22. $\quad \dfrac{x}{y^2} \div \dfrac{y}{x^2} = \dfrac{x}{y^2} \cdot \dfrac{x^2}{y} =$

$\quad \dfrac{x \cdot x \cdot x}{y \cdot y \cdot y} = \dfrac{x^3}{y^3}$

23. $\quad \dfrac{ab^2}{cd} \div \dfrac{c^2d^2}{ab^2} = \dfrac{ab^2}{cd} \cdot \dfrac{ab^2}{c^2d^2} =$

$\quad \dfrac{a \cdot b \cdot b \cdot a \cdot b \cdot b}{c \cdot d \cdot c \cdot c \cdot d \cdot d} = \dfrac{a^2b^4}{c^3d^3}$

24. $\quad \dfrac{a^2}{b^2} \div \dfrac{a}{b} = \dfrac{a^2}{b^2} \cdot \dfrac{b}{a} =$

$\quad \dfrac{a \cdot a \cdot b}{b \cdot b \cdot a} = \dfrac{a}{b}$

25. $\quad 3\dfrac{3}{5} \div 2\dfrac{7}{10} = -\dfrac{18}{5} \div \dfrac{27}{10} =$

$\quad \dfrac{\cancel{18}^2}{\cancel{5}} \cdot \dfrac{\cancel{10}^2}{\cancel{27}_3} = \dfrac{4}{3} = 1\dfrac{1}{3}$

26. $\quad (-1\dfrac{4}{7}) \div (7\dfrac{1}{3}) = -\dfrac{11}{7} \div \dfrac{22}{3} =$

$\quad -\dfrac{\cancel{11}}{7} \cdot \dfrac{3}{\cancel{22}_2} = -\dfrac{3}{14}$

27. $\quad \dfrac{3\dfrac{3}{4}}{-2\dfrac{5}{8}} = 3\dfrac{3}{4} \div (-2\dfrac{5}{8}) =$

$\quad \dfrac{15}{4} \div (-\dfrac{21}{8}) = \dfrac{\cancel{15}^5}{\cancel{4}} \cdot (-\dfrac{\cancel{8}^2}{\cancel{21}_7}) =$

$\quad -\dfrac{10}{7} = -1\dfrac{3}{7}$

28. $\quad (-1\dfrac{5}{7}) \div 24 = -\dfrac{12}{7} \div \dfrac{24}{1} =$

$\quad -\dfrac{\cancel{12}}{7} \cdot \dfrac{1}{\cancel{24}_2} = -\dfrac{1}{14}$

29. $\quad \dfrac{7}{8} \div \dfrac{7}{16} = \dfrac{\cancel{7}}{\cancel{8}} \cdot \dfrac{\cancel{16}^2}{\cancel{7}} =$

$\quad \dfrac{2}{1} = 2$

30. $\quad \dfrac{5}{6} \div \dfrac{15}{36} = \dfrac{\cancel{5}^1}{\cancel{6}_1} \cdot \dfrac{\cancel{36}^{6^2}}{\cancel{15}_3} =$

$\quad \dfrac{2}{1} = 2$

31. $\quad \dfrac{4}{5} \div \dfrac{16}{25} = \dfrac{\cancel{4}^1}{\cancel{5}_1} \cdot \dfrac{\cancel{25}^5}{\cancel{16}_4} = \dfrac{5}{4} = 1\dfrac{1}{4}$

32. $\quad 3\dfrac{3}{4} \div \dfrac{3}{8} = \dfrac{15}{4} \div \dfrac{3}{8} =$

$\quad \dfrac{\cancel{15}^5}{\cancel{4}} \cdot \dfrac{\cancel{8}^2}{\cancel{3}_1} = \dfrac{10}{1} = 10$ pillows

33. $\quad 84 \div 2\dfrac{2}{5} = \dfrac{84}{1} \div \dfrac{12}{5} =$

$\quad \dfrac{\cancel{84}^7}{1} \cdot \dfrac{5}{\cancel{12}} = \dfrac{35}{1} = 35$ shirts

34. $15 \div \dfrac{5}{8} = \dfrac{15}{1} \div \dfrac{5}{8} =$

$\dfrac{15^3}{1} \cdot \dfrac{8}{5} = \dfrac{24}{1} = 24$ glasses of pop

35. $1\dfrac{2}{3} \div \dfrac{1}{6} = \dfrac{5}{3} \cdot \dfrac{6^2}{1} =$

$\dfrac{10}{1} = 10 \dfrac{1}{6}$ cup measures

36. $3\dfrac{1}{8} \div \dfrac{5}{8} = \dfrac{25}{8} \div \dfrac{5}{8} =$

$\dfrac{25^5}{8_1} \cdot \dfrac{8^1}{5_1} = \dfrac{5}{1} = 5$ pillows

37. $100 \div 3\dfrac{1}{8} = \dfrac{100}{1} \div \dfrac{25}{8} =$

$\dfrac{100^4}{1} \cdot \dfrac{8}{25_1} = \dfrac{32}{1} = 32$ shirts

38. $12 \div \dfrac{4}{5} = \dfrac{12^3}{1} \cdot \dfrac{5}{4} =$

$\dfrac{15}{1} = 15$ glasses of pop

39. $2\dfrac{3}{4} \div \dfrac{1}{8} = \dfrac{11}{4} \div \dfrac{1}{8} =$

$\dfrac{11}{4} \cdot \dfrac{8^2}{1} = \dfrac{22}{1} = 22 \dfrac{1}{8}$ cup measures

40. If $\dfrac{4}{5}$ is painted there is $\dfrac{1}{5}$ left to paint.

$20 \div 4 = 5$ hours

5.5 Addition and Subtraction of Fractions and Mixed Numbers

1. $\dfrac{2}{5} + \dfrac{3}{5} = \dfrac{5}{5} = 1$

2. $\dfrac{4}{6} + \dfrac{1}{6} = \dfrac{5}{6}$

3. $\dfrac{6}{7} - \dfrac{3}{7} = \dfrac{3}{7}$

4. $\dfrac{3}{4} - \dfrac{1}{4} = \dfrac{2}{4} = \dfrac{1}{2}$

5. $\dfrac{1}{3} - \dfrac{2}{3} = \dfrac{1-2}{3} = \dfrac{1+(-2)}{3} = -\dfrac{1}{3}$

6. $\dfrac{-4}{9} - \dfrac{2}{9} = \dfrac{-4-2}{9} =$

$\dfrac{-4+(-2)}{9} = \dfrac{-6}{9} = \dfrac{-2}{3} = -\dfrac{2}{3}$

7. $\dfrac{1}{6} + \dfrac{1}{3} = \dfrac{1}{6} + \dfrac{2}{6} = \dfrac{3}{6} = \dfrac{1}{2}$

8. $\dfrac{2}{5} + \dfrac{7}{10} = \dfrac{4}{10} + \dfrac{7}{10} = \dfrac{11}{10}$ or $1\dfrac{1}{10}$

9. $\dfrac{2}{5} - \dfrac{3}{7} = \dfrac{14}{35} - \dfrac{15}{35} =$

$\dfrac{14+(-15)}{35} = \dfrac{-1}{35}$

10. $\dfrac{6}{5} - \dfrac{9}{15} = \dfrac{18}{15} - \dfrac{9}{15} = \dfrac{9}{15} = \dfrac{3}{5}$

11. $\dfrac{4}{9} + \dfrac{1}{3} = \dfrac{4}{9} + \dfrac{3}{9} = \dfrac{7}{9}$

12. $\dfrac{3}{10} + \dfrac{2}{5} = \dfrac{3}{10} + \dfrac{4}{10} = \dfrac{7}{10}$

41

13. $\quad \dfrac{5}{6} - \dfrac{1}{12} = \dfrac{10}{12} - \dfrac{1}{12} = \dfrac{9}{12} = \dfrac{3}{4}$

14. $\quad \dfrac{2}{3} + \dfrac{1}{4} = \dfrac{8}{12} + \dfrac{3}{12} = \dfrac{11}{12}$

15. $\quad \dfrac{2}{3} - \left(-\dfrac{1}{3}\right) = \dfrac{2}{3} + \dfrac{1}{3} = \dfrac{3}{3} = 1$

16. $\quad -2 - \dfrac{2}{3} = -\dfrac{6}{3} - \dfrac{2}{3} =$

$\quad -\dfrac{6}{3} + \left(-\dfrac{2}{3}\right) = -\dfrac{8}{3}$ or $-2\dfrac{2}{3}$

17. $\quad \dfrac{4}{3} + \dfrac{5}{2} = \dfrac{8}{6} + \dfrac{15}{6} = \dfrac{23}{6} = 3\dfrac{5}{6}$

18. $\quad -\dfrac{3}{10} + \dfrac{4}{15} = -\dfrac{9}{30} + \dfrac{8}{30} = -\dfrac{1}{30}$

19. $\quad \dfrac{4}{5} + \left(-\dfrac{5}{6}\right) = \dfrac{24}{30} + \left(-\dfrac{25}{30}\right) = -\dfrac{1}{30}$

20. $\quad \dfrac{7}{3} - \dfrac{13}{4} = \dfrac{28}{12} - \dfrac{39}{12} = -\dfrac{11}{12}$

21. $\quad \dfrac{5}{8} - \dfrac{11}{16} = \dfrac{10}{16} - \dfrac{11}{16} = -\dfrac{1}{16}$

22. $\quad -\dfrac{5}{9} - \dfrac{5}{6} = -\dfrac{10}{18} - \dfrac{15}{18} =$

$\quad -\dfrac{25}{18} = -1\dfrac{7}{18}$

23. $\quad \dfrac{10}{3} - \dfrac{7}{6} = \dfrac{20}{6} - \dfrac{7}{6} = \dfrac{13}{6} = 2\dfrac{1}{6}$

24. $\quad \dfrac{13}{8} - \dfrac{17}{6} = \dfrac{39}{24} - \dfrac{68}{24} =$

$\quad -\dfrac{29}{24} = -1\dfrac{5}{24}$

25. $\quad -\dfrac{9}{4} + \left(-\dfrac{14}{3}\right) = -\dfrac{27}{12} + \left(-\dfrac{56}{12}\right) =$

$\quad -\dfrac{83}{12} = -6\dfrac{11}{12}$

26. $\quad \dfrac{11}{28} + \dfrac{5}{21} = \dfrac{33}{84} + \dfrac{20}{84} = \dfrac{53}{84}$

27. $\quad \dfrac{-4}{3} - \left(-\dfrac{4}{5}\right) = -\dfrac{20}{15} - \left(-\dfrac{12}{15}\right) =$

$\quad \dfrac{-20}{15} + \dfrac{12}{15} = \dfrac{-8}{15}$

28. $\quad \dfrac{7}{8} - \dfrac{5}{7} = \dfrac{49}{56} - \dfrac{40}{56} = \dfrac{9}{56}$

29. $\quad \dfrac{9}{14} + \left(-\dfrac{5}{21}\right) = \dfrac{27}{42} + \left(-\dfrac{10}{42}\right) = \dfrac{17}{42}$

30. $\quad \dfrac{9}{4} + \left(-\dfrac{3}{2}\right) = \dfrac{9}{4} + \left(-\dfrac{6}{4}\right) = \dfrac{3}{4}$

31. $\quad \dfrac{23}{5} + \left(-\dfrac{4}{15}\right) = \dfrac{69}{15} + \left(-\dfrac{4}{15}\right) =$

$\quad \dfrac{65}{15} = \dfrac{13}{3} = 4\dfrac{1}{3}$

32. $\quad \dfrac{4}{15} + \left(-\dfrac{17}{45}\right) = \dfrac{12}{45} + \left(-\dfrac{17}{45}\right) =$

$\quad -\dfrac{5}{45} = -\dfrac{1}{9}$

33. $\quad \dfrac{2}{3} + \dfrac{4}{5} + \dfrac{11}{15} = \dfrac{10}{15} + \dfrac{12}{15} + \dfrac{11}{15} =$

$\quad \dfrac{33}{15} = \dfrac{11}{5} = 2\dfrac{1}{5}$

34. $\dfrac{5}{6} + \dfrac{7}{12} + \dfrac{1}{24} =$

$\dfrac{20}{24} + \dfrac{14}{24} + \dfrac{1}{24} = \dfrac{35}{24}$ or $1\dfrac{11}{24}$

35. $\dfrac{9}{10} - \dfrac{2}{15} = \dfrac{27}{30} - \dfrac{4}{30} = \dfrac{23}{30}$

36. $\dfrac{5}{6} - \dfrac{2}{9} = \dfrac{15}{18} - \dfrac{4}{18} = \dfrac{11}{18}$

37. $\dfrac{3}{4} + \dfrac{3}{8} = \dfrac{6}{8} + \dfrac{3}{8} =$

$\dfrac{9}{8} = 1\dfrac{1}{8}$ cups of sugar

38. $\dfrac{1}{2} + \dfrac{3}{4} + \dfrac{2}{3} + \dfrac{5}{6} + \dfrac{3}{4}$

$\dfrac{6}{12} + \dfrac{9}{12} + \dfrac{8}{12} + \dfrac{10}{12} + \dfrac{9}{12} =$

$\dfrac{42}{12} = \dfrac{7}{2} = 3\dfrac{1}{2}$ miles

39. $\dfrac{3}{10} + \dfrac{2}{5} = \dfrac{3}{10} + \dfrac{4}{10} =$

$\dfrac{7}{10}$ spent

$\dfrac{10}{10} - \dfrac{7}{10} = \dfrac{3}{10}$ left.

40. $\dfrac{7}{8} + \dfrac{4}{5} = \dfrac{35}{40} + \dfrac{32}{40} =$

$\dfrac{67}{40} = 1\dfrac{27}{40}$ cups.

41. $\dfrac{7}{8} + \dfrac{2}{3} + \dfrac{5}{8} + \dfrac{3}{4} + \dfrac{2}{3}$

$\dfrac{21}{24} + \dfrac{16}{24} + \dfrac{15}{24} + \dfrac{18}{24} + \dfrac{16}{24}$

$\dfrac{86}{24} = \dfrac{43}{12} = 3\dfrac{7}{12}$

42. $\dfrac{1}{3} + \dfrac{1}{6} = \dfrac{2}{6} + \dfrac{1}{6} =$

$\dfrac{3}{6} = \dfrac{1}{2}$ spent

$\dfrac{2}{2} - \dfrac{1}{2} = \dfrac{1}{2}$ left.

43. $\dfrac{5}{8} + \dfrac{3}{8} + \dfrac{5}{8} + \dfrac{7}{8} + 1\dfrac{1}{8} =$

$\dfrac{5}{8} + \dfrac{3}{8} + \dfrac{5}{8} + \dfrac{7}{8} + \dfrac{9}{8} =$

$\dfrac{29}{8} = 3\dfrac{5}{8}$ points rise for the week.

44. $\dfrac{1}{3} + \dfrac{1}{4} + \dfrac{3}{8} = \dfrac{8}{24} + \dfrac{6}{24} + \dfrac{9}{24}$

$\dfrac{23}{24}$ scored in first 3 quarters

$\dfrac{24}{24} - \dfrac{23}{24} =$

$\dfrac{1}{24}$ scored in final quarter.

45. $1\frac{2}{3} + \frac{8}{9} + 1\frac{1}{9}$

$$1\frac{2}{3} = 1\frac{6}{9}$$
$$\frac{8}{9} = \frac{8}{9}$$
$$+\ 1\frac{1}{9} = 1\frac{1}{9}$$
$$\overline{\phantom{+\ 1\frac{1}{9} = 1\frac{1}{9}}}$$
$$3\frac{6}{9} =$$

$3\frac{2}{3}$ inches in the perimeter.

46. $4\frac{3}{4} + 2\frac{3}{8} + 4\frac{3}{4} + 2\frac{3}{8}$

$$4\frac{3}{4} = 4\frac{6}{8}$$
$$2\frac{3}{8} = 2\frac{3}{8}$$
$$4\frac{3}{4} = 4\frac{6}{8}$$
$$+\ 2\frac{3}{8} = 2\frac{3}{8}$$
$$\overline{\phantom{+\ 2\frac{3}{8} = 2\frac{3}{8}}}$$
$$14\frac{2}{8} =$$

$14\frac{1}{4}$ inches in the perimeter.

5.6 Order of Operations and Complex Fractions

1. $\dfrac{\frac{7}{8}}{\frac{3}{4}} = \dfrac{8}{8} \cdot \dfrac{\left(\frac{7}{8}\right)}{\left(\frac{3}{4}\right)} = \dfrac{7}{6}$ or $1\frac{1}{6}$

2. $\dfrac{\frac{3}{10}}{\frac{1}{2}} = \dfrac{3}{10} \div \dfrac{1}{2} = \dfrac{3}{10_5} \cdot \dfrac{2^1}{1} = \dfrac{3}{5}$

3. $\dfrac{\frac{1}{3}}{\frac{5}{6}} = \dfrac{6}{6} \cdot \dfrac{\left(\frac{1}{3}\right)}{\left(\frac{5}{6}\right)} = \dfrac{2}{5}$

4. $\dfrac{\frac{3}{4}}{\frac{2}{3}} = \dfrac{3}{4} \div \dfrac{2}{3} = \dfrac{3}{4} \cdot \dfrac{3}{2} =$

$\dfrac{9}{8} = 1\frac{1}{8}$

5. $\dfrac{2}{3} + 4\left(\dfrac{1}{2} + \dfrac{3}{2}\right) = \dfrac{2}{3} + 4\left(\dfrac{4}{2}\right) =$

$\dfrac{2}{3} + 4(2) = \dfrac{2}{3} + 8 = 8\frac{2}{3}$

6. $5 + 12\left(\dfrac{1}{6} + \dfrac{1}{8}\right) =$

$5 + 12\left(\dfrac{4}{24} + \dfrac{3}{24}\right) = 5 + 12\left(\dfrac{7}{24}\right) =$

$5 + \dfrac{7}{2} = 5 + 3\dfrac{1}{2} = 8\frac{1}{2}$

7. $\left(\dfrac{2}{3} + \dfrac{2}{9}\right)\left(\dfrac{1}{2} - \dfrac{2}{5}\right) =$

$\left(\dfrac{6}{9} + \dfrac{2}{9}\right)\left(\dfrac{5}{10} - \dfrac{4}{10}\right) =$

$\dfrac{8^4}{9} \cdot \dfrac{1}{10_5} = \dfrac{4}{45}$

8. $(\frac{3}{4} - \frac{4}{5})(\frac{3}{7} - \frac{1}{14}) =$

$(\frac{15}{20} - \frac{16}{20})(\frac{6}{14} - \frac{1}{14}) =$

$(-\frac{1}{20}_4)(\frac{5}{14}^1) = -\frac{1}{56}$

9. $\frac{4}{7} \div (-\frac{2}{7})(-\frac{3}{8}) =$

$\frac{4}{7}_1 \cdot (-\frac{7}{2}^1)(-\frac{3}{8}_2) = \frac{3}{4}$

10. $\frac{5}{9} \div (-\frac{35}{39})(\frac{14}{13}) =$

$\frac{5}{9}_3^1 \cdot (-\frac{39}{35}^{3^1}_7)(\frac{14}{13}^2_1) = -\frac{2}{3}$

11. $(\frac{3}{4})(\frac{1}{2})^2 + \frac{1}{8} = (\frac{3}{4})(\frac{1}{2})(\frac{1}{2}) + \frac{1}{8}$

$\frac{3}{16} + \frac{1}{8} = \frac{3}{16} + \frac{2}{16} = \frac{5}{16}$

12. $(2\frac{1}{2})^2 + (1\frac{1}{2})^2 = (\frac{5}{2})^2 + (\frac{3}{2})^2 =$

$\frac{5}{2} \cdot \frac{5}{2} + \frac{3}{2} \cdot \frac{3}{2} = \frac{25}{4} + \frac{9}{4} =$

$\frac{34}{4} = \frac{17}{2} = 8\frac{1}{2}$

13. $(\frac{2}{3})^2(\frac{3}{4}) - \frac{2}{5} =$

$\frac{2}{3}^1 \cdot \frac{2}{3}^1 \cdot \frac{3}{4}^1 - \frac{2}{5} = \frac{1}{3} - \frac{2}{5}$

$\frac{5}{15} - \frac{6}{15} = -\frac{1}{15}$

14. $\frac{2}{3}(\frac{3}{4} + 3) + (\frac{1}{3})^2 =$

$\frac{2}{3}(\frac{3}{4} + \frac{12}{4}) + (\frac{1}{3})(\frac{1}{3}) =$

$\frac{2}{3}(\frac{15}{4}^5_2) + \frac{1}{9} = \frac{5}{2} + \frac{1}{9} =$

$\frac{45}{18} + \frac{2}{18} = \frac{47}{18} = 2\frac{11}{18}$

15. $\frac{7}{8} + \frac{2}{3}(\frac{1}{2} + \frac{6}{5}) =$

$\frac{7}{8} + \frac{2}{3}^1(\frac{17}{10}_5) = \frac{7}{8} + \frac{17}{15} =$

$\frac{105}{120} + \frac{136}{120} = \frac{241}{120} = 2\frac{1}{120}$

16. $\frac{\frac{15}{6}}{9} = \frac{15}{6} \div 9 = \frac{15}{6}^5 \cdot \frac{1}{9}_3 = \frac{5}{18}$

17. $\frac{\frac{2}{5} + \frac{1}{6}}{\frac{3}{4} - \frac{7}{8}} = (\frac{2}{5} + \frac{1}{6}) \div (\frac{3}{4} - \frac{7}{8})$

$(\frac{12}{30} + \frac{5}{30}) \div (\frac{6}{8} - \frac{7}{8}) =$

$(\frac{17}{30}) \div (-\frac{1}{8}) = \frac{17}{30}_{15} \cdot (-\frac{8}{1}^4) =$

$-\frac{68}{15} = -4\frac{8}{15}$

18. $$\frac{\dfrac{1}{8}+\dfrac{3}{4}}{\dfrac{1}{2}-\dfrac{1}{3}} = \frac{\left(\dfrac{1}{8}+\dfrac{3}{4}\right)}{\left(\dfrac{1}{2}-\dfrac{1}{3}\right)} \cdot \frac{24}{24} =$$

$$\frac{\cancel{24}^{3}\left(\dfrac{1}{\cancel{8}}\right)+\cancel{24}^{6}\left(\dfrac{3}{\cancel{4}}\right)}{\cancel{24}^{12}\left(\dfrac{1}{\cancel{2}}\right)-\cancel{24}^{8}\left(\dfrac{1}{\cancel{3}}\right)} =$$

$$\frac{3+18}{12-8} = \frac{21}{4} \quad \text{or} \quad 5\frac{1}{4}$$

19. $$\frac{\dfrac{3}{16}+5}{6-\dfrac{7}{8}} = \left(\dfrac{3}{16}+5\right) \div \left(6-\dfrac{7}{8}\right) =$$

$$\left(\frac{3}{16}+\frac{80}{16}\right) \div \left(\frac{48}{8}-\frac{7}{8}\right) =$$

$$\frac{83}{16} \div \frac{41}{8} = \frac{83}{\cancel{16}_{2}} \cdot \frac{\cancel{8}^{1}}{41} = \frac{83}{82}$$

$$\text{or} \quad 1\frac{1}{82}$$

20. $$\frac{\dfrac{-4}{3}+\dfrac{5}{6}}{\dfrac{-3}{4}+\dfrac{2}{9}} = \frac{\left(\dfrac{-4}{3}+\dfrac{5}{6}\right)}{\left(\dfrac{-3}{4}+\dfrac{2}{9}\right)} \cdot \frac{36}{36} =$$

$$\frac{\cancel{36}^{12}\left(\dfrac{-4}{\cancel{3}}\right)+\cancel{36}^{6}\left(\dfrac{5}{\cancel{6}}\right)}{\cancel{36}^{9}\left(\dfrac{-3}{\cancel{4}}\right)+\cancel{36}^{4}\left(\dfrac{2}{\cancel{9}}\right)} = \frac{-48+30}{-27+8} =$$

$$\frac{-18}{-19} = \frac{18}{19}$$

21. $$\frac{\dfrac{2}{3}+\dfrac{3}{2}}{\dfrac{1}{3}-\dfrac{5}{6}} = \frac{\left(\dfrac{2}{3}+\dfrac{3}{2}\right)}{\left(\dfrac{1}{3}-\dfrac{5}{6}\right)} \cdot \frac{6}{6} =$$

$$\frac{\cancel{6}^{2}\left(\dfrac{2}{\cancel{3}}\right)+\cancel{6}^{3}\left(\dfrac{3}{\cancel{2}}\right)}{\cancel{6}_{2}\left(\dfrac{1}{\cancel{3}}\right)-\cancel{6}\left(\dfrac{5}{\cancel{6}}\right)} = \frac{4+9}{2-5} =$$

$$\frac{13}{-3} \quad \text{or} \quad -4\frac{1}{3}$$

22. $$\frac{\dfrac{3}{5}-\dfrac{7}{10}}{\dfrac{3}{20}-\dfrac{1}{5}} = \frac{\left(\dfrac{3}{5}-\dfrac{7}{10}\right)}{\left(\dfrac{3}{20}-\dfrac{1}{5}\right)} \cdot \frac{20}{20} =$$

$$\frac{\cancel{20}^{4}\left(\dfrac{3}{\cancel{5}}\right)-\cancel{20}^{2}\left(\dfrac{7}{\cancel{10}}\right)}{\cancel{20}\left(\dfrac{3}{\cancel{20}}\right)-\cancel{20}_{4}\left(\dfrac{1}{\cancel{5}}\right)} = \frac{12-14}{3-4} =$$

$$\frac{-2}{-1} = 2$$

23. $$\frac{5-\dfrac{3}{4}}{2-\dfrac{1}{3}} = \left(5-\dfrac{3}{4}\right) \div \left(2-\dfrac{1}{3}\right) =$$

$$\left(\frac{20}{4}-\frac{3}{4}\right) \div \left(\frac{6}{3}-\frac{1}{3}\right) = \frac{17}{4} \div \frac{5}{3} =$$

$$\frac{17}{4} \cdot \frac{3}{5} = \frac{51}{20} \quad \text{or} \quad 2\frac{11}{20}$$

24.

$$\frac{\frac{4}{3} - \frac{5}{6}}{\frac{3}{4} + \frac{2}{5}} = \left(\frac{4}{3} - \frac{5}{6}\right) \div \left(\frac{3}{4} + \frac{2}{5}\right) =$$

$$\left(\frac{8}{6} - \frac{5}{6}\right) \div \left(\frac{15}{20} + \frac{8}{20}\right) = \frac{3}{6} \div \frac{23}{20} =$$

$$\frac{\cancel{3}}{\cancel{6}_2} \cdot \frac{\cancel{20}^{10}}{23} = \frac{10}{23}$$

25.

$$\frac{\frac{4}{3} + \frac{5}{6}}{\frac{2}{3} - \frac{3}{2}} = \frac{\frac{4}{3} + \frac{5}{6}}{\frac{2}{3} - \frac{3}{2}} \cdot \frac{6}{6} =$$

$$\frac{\cancel{6}^2\left(\frac{4}{\cancel{3}}\right) + \cancel{6}\left(\frac{5}{\cancel{6}}\right)}{\cancel{6}_2\left(\frac{2}{\cancel{3}}\right) - \cancel{6}_3\left(\frac{3}{\cancel{2}}\right)} = \frac{8 + 5}{4 - 9} =$$

$$\frac{13}{-5} \text{ or } -2\frac{3}{5}$$

26.

$$\frac{\frac{13}{18} - \frac{11}{24}}{\frac{5}{12} - \frac{7}{36}} = \frac{\frac{13}{18} - \frac{11}{24}}{\frac{5}{12} - \frac{7}{36}} =$$

$$\frac{\cancel{72}^4\left(\frac{13}{\cancel{18}}\right) - \cancel{72}^3\left(\frac{11}{\cancel{24}}\right)}{\cancel{72}_6\left(\frac{5}{\cancel{12}}\right) - \cancel{72}_2\left(\frac{7}{\cancel{36}}\right)} = \frac{52 - 33}{30 - 14} =$$

$$\frac{19}{16} \text{ or } 1\frac{3}{16}$$

27.

$$\frac{\frac{2}{5} + \frac{7}{2}}{\frac{12}{3} - \frac{6}{5}} = \left(\frac{2}{5} + \frac{7}{2}\right) \div \left(\frac{12}{3} - \frac{6}{5}\right) =$$

$$\left(\frac{4}{10} + \frac{35}{10}\right) \div \left(\frac{60}{15} - \frac{18}{15}\right) =$$

$$\frac{39}{10} \div \frac{42}{15} = \frac{39}{\cancel{10}_2} \cdot \frac{\cancel{15}^5}{\cancel{42}_{14}} = \frac{39}{28}$$

$$\text{or } 1\frac{11}{28}$$

5.7 Solving Equations Using the Multiplication Property of Equality

1.
$$3x = 15$$
$$\frac{1}{3} \cdot 3x = 15 \cdot \frac{1}{3}$$
$$x = 5$$

2.
$$2x = 28$$
$$\frac{1}{2} \cdot 2x = 28 \cdot \frac{1}{2}$$
$$x = 14$$

3.
$$5q = 25$$
$$\frac{1}{5} \cdot 5q = 25 \cdot \frac{1}{5}$$
$$q = 5$$

4.
$$7r = 35$$
$$\frac{1}{7} \cdot 7r = 35 \cdot \frac{1}{7}$$
$$r = 5$$

5.
$$60 = 6z$$
$$\frac{1}{6} \cdot 60 = 6z \cdot \frac{1}{6}$$
$$10 = z$$

6.
$$42 = 6y$$
$$\frac{1}{6} \cdot 42 = 6y \cdot \frac{1}{6}$$
$$7 = y$$

7.
$$-2x = 14$$
$$-\frac{1}{2} \cdot (-2x) = (14) \cdot -\frac{1}{2}$$
$$x = -7$$

8.
$$-3x = 21$$
$$-\frac{1}{3} \cdot (-3x) = 21 \cdot -\frac{1}{3}$$
$$x = -7$$

9.
$$8y = -64$$
$$\frac{8y}{8} = \frac{-64}{8}$$
$$y = -8$$

10.
$$9y = -63$$
$$\frac{9y}{9} = \frac{-63}{9}$$
$$y = -7$$

11.
$$-4m = 12$$
$$\frac{-4m}{-4} = \frac{12}{-4}$$
$$m = -3$$

12.
$$-3n = 18$$
$$\frac{-3n}{-3} = \frac{18}{-3}$$
$$n = -6$$

13.
$$6s = -24$$
$$\frac{6s}{6} = \frac{-24}{6}$$
$$s = -4$$

14.
$$7t = -28$$
$$\frac{7t}{7} = \frac{-28}{7}$$
$$t = -4$$

15.
$$-2z = -26$$
$$\frac{-2z}{-2} = \frac{-26}{-2}$$
$$z = 13$$

16.
$$-5w = -45$$
$$\frac{-5w}{-5} = \frac{-45}{-5}$$
$$w = 9$$

17.
$$\frac{1}{3}x = 4$$
$$\frac{3}{1} \cdot \frac{1}{3} = 4 \cdot \frac{3}{1}$$
$$x = 12$$

18.
$$\frac{1}{2}x = 5$$
$$\frac{2}{1} \cdot \frac{1}{2} = 5 \cdot \frac{2}{1}$$
$$x = 10$$

19.
$$-\frac{1}{4}y = 2$$
$$\left(-\frac{4}{1}\right)\left(-\frac{1}{4}y\right) = 2 \cdot \left(-\frac{4}{1}\right)$$
$$y = -8$$

20.
$$-\frac{1}{5}y = 3$$
$$-\frac{5}{1} \cdot \left(-\frac{1}{5}y\right) = 3 \cdot \left(-\frac{5}{1}\right)$$
$$y = -15$$

21.
$$\frac{2}{7}x = 8$$
$$\frac{7}{2} \cdot \frac{2}{7}x = 8 \cdot \frac{7}{2}$$
$$x = 28$$

22.
$$\frac{3}{5}x = 9$$
$$\frac{5}{3}\left(\frac{3}{5}x\right) = 9\left(\frac{5}{3}\right)$$
$$x = 15$$

23.
$$-\frac{3}{11}y = 12$$
$$-\frac{11}{3}\left(-\frac{3}{1}y\right) = 12\left(-\frac{11}{3}\right)$$
$$y = -44$$

24.
$$-\frac{2}{13}y = 26$$
$$-\frac{13}{2}\left(-\frac{2}{13}y\right) = 26\left(-\frac{13}{2}\right)$$
$$y = -169$$

25.
$$\frac{3}{4}x = 27$$
$$\frac{4}{3}\left(\frac{3}{4}x\right) = 27\left(\frac{4}{3}\right)$$
$$x = 36$$

26.
$$\frac{2}{5}x = 32$$
$$\frac{5}{2}\left(\frac{2}{5}x\right) = 32 \cdot \frac{5}{2}$$
$$x = 80$$

27.
$$7x = \frac{29}{14}$$
$$\frac{1}{7} \cdot 7x = \frac{29}{14} \cdot \frac{1}{7}$$
$$x = \frac{29}{98}$$

28.
$$9x = \frac{19}{10}$$
$$\frac{1}{9} \cdot 9x = \frac{19}{10} \cdot \frac{1}{9}$$
$$x = \frac{19}{90}$$

29.
$$\frac{8}{5}y = -31$$
$$\frac{5}{8} \cdot \frac{8}{5}y = -31 \cdot \frac{5}{8}$$
$$y = \frac{-155}{8} = -19\frac{3}{8}$$

30.
$$\frac{16}{7}y = -22$$
$$\frac{7}{16}\left(\frac{16}{7}y\right) = -22 \cdot \frac{7}{16}$$
$$y = \frac{-77}{8} = -9\frac{5}{8}$$

31.

$$3x = 687$$
$$\frac{3x}{3} = \frac{687}{3}$$
$$x = 229$$

Burn 229 calories running 1 mile.

32.

$$2x = 5$$
$$\frac{1}{2} \cdot (2x) = 5 \cdot \frac{1}{2}$$
$$x = \frac{5}{2} = 2\frac{1}{2}$$

It takes 2½ minutes to walk from the dorm to class.

REVIEW

1. See Section 5.1
2. See Section 5.1
3. See Section 5.1
4. See Section 5.1
5. See Section 5.1
6. See Section 5.1
7. See Section 5.2
8. See Section 5.4
9. See Section 5.4
10. See Section 5.5
11. See Section 5.6
12. See Section 5.7
13. See Section 5.7
14. 1 numerator   6 denominator
15. 23 numerator   4 denominator
16. 7 numerator   9 denominator
17. 9 numerator   8 denominator
18. 12 numerator   7 denominator
19. 0 numerator   5 denominator
20. 7 numerator   1 denominator
21. 1 numerator   12 denominator
22. 5 numerator   2 denominator
23. 37 numerator   86 denominator

24. $\frac{1}{2} = \frac{?}{18} = \frac{1}{2} \cdot \frac{9}{9} = \frac{9}{18}$

25. $\frac{3}{8} = \frac{?}{24} \qquad \frac{3}{8} \cdot \frac{3}{3} = \frac{9}{24}$

26. $\frac{2}{9} = \frac{?}{24} \qquad \frac{2}{9} \cdot \frac{6}{6} = \frac{12}{54}$

27. $\frac{3}{4} = \frac{?}{8x} \qquad \frac{3}{4} \cdot \frac{2x}{2x} = \frac{6x}{8x}$

28. $\frac{15}{36} = \frac{?}{12} \qquad \frac{15}{36} \div \frac{3}{3} = \frac{5}{12}$

29. $\frac{3}{9} = \frac{?}{3} \qquad \frac{3}{9} \div \frac{3}{3} = \frac{1}{3}$

30. $\frac{35}{30} = \frac{?}{6} \qquad \frac{35}{30} \div \frac{5}{5} = \frac{7}{6}$

31. $\frac{36}{72} = \frac{?}{2} \qquad \frac{36}{72} \div \frac{36}{36} = \frac{1}{2}$

32. $\frac{55}{110} = \frac{?}{10} \qquad \frac{55}{110} \div \frac{11}{11} = \frac{5}{10}$

33. $\frac{55}{110} = \frac{?}{2} \qquad \frac{55}{110} \div \frac{55}{55} = \frac{1}{2}$

34. $\frac{19}{38} = \frac{19}{2 \cdot 19} = \frac{1}{2}$

35. $\frac{12}{36} = \frac{2 \cdot 2 \cdot 3}{2 \cdot 2 \cdot 3 \cdot 3} = \frac{1}{3}$

36. $\frac{30a^2b}{40ab^2} =$

$$\frac{2 \cdot 3 \cdot 5 \cdot a \cdot a \cdot b}{2 \cdot 2 \cdot 2 \cdot 5 \cdot a \cdot b \cdot b} = \frac{3a}{4b}$$

37. $\dfrac{15x}{16y} = \dfrac{3 \cdot 5 \cdot x}{2 \cdot 2 \cdot 2 \cdot 2 \cdot y} = \dfrac{15x}{16y}$

38. $\dfrac{72x^3}{9x^2} =$

$\dfrac{2 \cdot 2 \cdot 2 \cdot 3 \cdot 3 \cdot x \cdot x \cdot x}{3 \cdot 3 \cdot x \cdot x} = \dfrac{8x}{1} = 8x$

39. $\dfrac{4x^2}{8x} = \dfrac{2 \cdot 2 \cdot x \cdot x}{2 \cdot 2 \cdot 2 \cdot x} = \dfrac{x}{2}$

40. $\dfrac{100}{200} = \dfrac{2 \cdot 2 \cdot 5 \cdot 5}{2 \cdot 2 \cdot 2 \cdot 5 \cdot 5} = \dfrac{1}{2}$

41. $\dfrac{35}{63} = \dfrac{5 \cdot 7}{7 \cdot 9} = \dfrac{5}{9}$

42. $\left(\dfrac{3}{4}\right)\left(-\dfrac{8}{9}\right)(-4) =$

$\dfrac{3 \cdot (-8) \cdot (-4)}{4 \cdot 9_3} = \dfrac{8}{3} = 2\dfrac{2}{3}$

43. $\left(5\dfrac{1}{2}\right)\left(3\dfrac{1}{3}\right) = \dfrac{11}{\cancel{2}} \cdot \dfrac{\cancel{10}^5}{3} = \dfrac{55}{3} = 18\dfrac{1}{3}$

44. $\left(2\dfrac{1}{2}\right)\left(4\dfrac{6}{7}\right) = \dfrac{5}{\cancel{2}} \cdot \dfrac{\cancel{34}^{17}}{7} = \dfrac{85}{7} = 12\dfrac{1}{7}$

45. $\left(\dfrac{20x}{30}\right)\left(\dfrac{15y}{88}\right) =$

$\dfrac{\cancel{20}x}{\cancel{30}_{\cancel{2}_1}} \cdot \dfrac{\cancel{15}^5 y}{\cancel{88}_{44}} = \dfrac{5xy}{44}$

46. $-\dfrac{5^1}{6_1} \cdot \dfrac{6^1}{5_1} = -\dfrac{1}{1} = -1$

47. $\left(\dfrac{9}{12}\right)\left(\dfrac{10}{15}\right) = \dfrac{\cancel{9}^{3^1} \cdot \cancel{10}^{2^1}}{\cancel{12}_{4_2} \cdot \cancel{15}_{3_1}} = \dfrac{1}{2}$

48. $\dfrac{x^2}{y} \cdot \dfrac{y^2}{x} = \dfrac{x \cdot x \cdot y \cdot y}{y \cdot x} = \dfrac{xy}{1} = xy$

49. $\dfrac{7}{8} \cdot \dfrac{8}{21} = \dfrac{7^1}{\cancel{8}_1} \cdot \dfrac{\cancel{8}^1}{\cancel{21}_3} = \dfrac{1}{3}$

50. $\left(-\dfrac{3}{18}\right)\left(-\dfrac{9}{10}\right)\left(\dfrac{5}{14}\right)\left(-\dfrac{8}{21}\right) =$

$\dfrac{(-3)(-9)(5)(-8)}{18 \cdot 10 \cdot 14 \cdot 21} = -\dfrac{1}{49}$

51. $\dfrac{9}{10} \cdot \dfrac{5}{18} = \dfrac{\cancel{9}^1}{\cancel{10}_2} \cdot \dfrac{\cancel{5}^1}{\cancel{18}_2} = \dfrac{1}{4}$

52. $\dfrac{2}{\cancel{4}} \cdot \cancel{2024} = \dfrac{3}{4_1} \cdot \dfrac{2024^{506}}{1} =$

$\dfrac{1518}{1} = 1518$

1518 student will graduate in 4 years.

53. $\left(6\dfrac{3}{7}\right)\left(4\dfrac{1}{11}\right) = \dfrac{45}{7} \cdot \dfrac{45}{11} =$

$\dfrac{2025}{77} = 26\dfrac{23}{77}$

54. $6^2 \cdot \dfrac{1}{\cancel{3}_1} = 2$ Price reduced by $2.

55. $\dfrac{2}{5} \div \dfrac{15}{4} = \dfrac{2}{5} \cdot \dfrac{4}{15} = \dfrac{8}{75}$

56. $\dfrac{9}{8} \div \dfrac{3}{4} = \dfrac{\cancel{9}^3}{\cancel{8}_2} \cdot \dfrac{\cancel{4}^1}{\cancel{3}_1} = \dfrac{3}{2} = 1\dfrac{1}{2}$

57. $-\dfrac{7}{8} \div \dfrac{24}{14} = \dfrac{7}{\cancel{8}_4} \cdot \dfrac{\cancel{14}^7}{24} = -\dfrac{49}{96}$

58. $\dfrac{2}{5} \div \dfrac{3}{5} = \dfrac{2}{\cancel{5}_1} \cdot \dfrac{\cancel{5}^1}{3} = \dfrac{2}{3}$

59. $-2\dfrac{4}{5} \div (-\dfrac{7}{10}) =$

$(-\dfrac{\cancel{14}^2}{\cancel{5}_1}) \cdot (-\dfrac{\cancel{10}^2}{\cancel{7}_1}) = 4$

60. $\dfrac{6}{7} \div \dfrac{3}{14} = \dfrac{\cancel{6}^2}{\cancel{7}_1} \cdot \dfrac{\cancel{14}^2}{\cancel{3}_1} = 4$

61. $\dfrac{1}{6} \div (-3\dfrac{2}{3}) = \dfrac{1}{\cancel{6}_2} \cdot (-\dfrac{\cancel{3}^1}{11}) = -\dfrac{1}{22}$

62. $\dfrac{1}{96} \div (-\dfrac{3}{50}) =$

$\dfrac{1}{\cancel{96}_{48}} \cdot (-\dfrac{\cancel{50}^{25}}{3}) = -\dfrac{25}{144}$

63. $\dfrac{2}{3} \div (-\dfrac{4}{21}) =$

$\dfrac{\cancel{2}^1}{\cancel{3}_1} \cdot (-\dfrac{\cancel{21}^7}{\cancel{4}_2}) = -\dfrac{7}{2} = -3\dfrac{1}{2}$

64. $7\dfrac{1}{2} \div \dfrac{2}{3} = \dfrac{15}{2} \cdot \dfrac{3}{2} = \dfrac{45}{4} = 11\dfrac{1}{4}$

11¼ servings of pop.

65. $\dfrac{8}{9} \div \dfrac{8}{3} = \dfrac{\cancel{8}^1}{\cancel{9}_3} \cdot \dfrac{\cancel{3}^1}{\cancel{8}_1} = \dfrac{1}{3}$

66. $\dfrac{7}{8} \div \dfrac{1}{16} = \dfrac{7}{\cancel{8}_1} \cdot \dfrac{\cancel{16}^2}{1} = 14$ omlettes

67. $\dfrac{5}{6} + \dfrac{5}{12} = \dfrac{10}{12} + \dfrac{5}{12} =$

$\dfrac{15}{12} = \dfrac{5}{4} = 1\dfrac{1}{4}$

68. $\dfrac{1}{3} + \dfrac{1}{4} + \dfrac{1}{12} =$

$\dfrac{4}{12} + \dfrac{3}{12} + \dfrac{1}{12} = \dfrac{8}{12} = \dfrac{2}{3}$

69. $-1\dfrac{1}{2} + 3\dfrac{1}{4} = -\dfrac{3}{2} + \dfrac{13}{4} =$

$-\dfrac{6}{4} + \dfrac{13}{4} = \dfrac{7}{4} = 1\dfrac{3}{4}$

70. $\dfrac{1}{2} + \dfrac{3}{4} + (-\dfrac{3}{16}) =$

$\dfrac{8}{16} + \dfrac{12}{16} + (-\dfrac{3}{16}) = \dfrac{17}{16} = 1\dfrac{1}{16}$

71. $\dfrac{5}{7} - \dfrac{4}{5} = \dfrac{25}{35} - \dfrac{28}{35} = \dfrac{-3}{35}$

72. $\dfrac{1}{6} + (-\dfrac{3}{8}) + \dfrac{1}{2} =$

$\dfrac{4}{24} + (-\dfrac{9}{24}) + \dfrac{12}{24} = \dfrac{7}{24}$

73. $-1\dfrac{3}{4} - \dfrac{11}{12} = -\dfrac{7}{4} + (-\dfrac{11}{12}) =$

$-\dfrac{21}{12} + (-\dfrac{11}{12}) = \dfrac{-32}{12} = \dfrac{-8}{3} = -2\dfrac{2}{3}$

52

74. $\dfrac{5}{12} + (-\dfrac{7}{16}) + (-\dfrac{1}{24}) =$

$\dfrac{20}{48} + (-\dfrac{21}{48}) + (-\dfrac{2}{48}) =$

$-\dfrac{3}{48} = -\dfrac{1}{16}$

75. $\dfrac{3}{4} + \dfrac{2}{3} = \dfrac{9}{12} + \dfrac{8}{12} = \dfrac{17}{12} = 1\dfrac{5}{12}$

76. $\dfrac{11}{20} + \dfrac{2}{5} + \dfrac{13}{40} =$

$\dfrac{22}{40} + \dfrac{16}{40} + \dfrac{13}{40} = \dfrac{51}{40} = 1\dfrac{11}{40}$

77. $\dfrac{9}{4} + \dfrac{5}{6} + \dfrac{1}{12} =$

$\dfrac{27}{12} + \dfrac{10}{12} + \dfrac{1}{12} = \dfrac{38}{12} = \dfrac{19}{6} = 3\dfrac{1}{6}$

78. $\dfrac{7}{3} - \dfrac{7}{4} = \dfrac{28}{12} - \dfrac{21}{12} = \dfrac{7}{12}$

79. $1 - \dfrac{2}{5} + \dfrac{4}{15} =$

$\dfrac{15}{15} - \dfrac{6}{15} + \dfrac{4}{15} = \dfrac{5}{15} = \dfrac{1}{3}$

⅓ of the family's income is left.

80. $\dfrac{1}{2} + \dfrac{1}{3} + \dfrac{1}{3} = \dfrac{3}{6} + \dfrac{2}{6} + \dfrac{2}{6}$

$\dfrac{7}{6} = 1\dfrac{1}{6}$

81. $\dfrac{1}{2} + \dfrac{3}{10} + \dfrac{3}{5} = \dfrac{5}{10} + \dfrac{3}{10} + \dfrac{6}{10}$

$\dfrac{14}{10} = \dfrac{7}{5} = 1\dfrac{2}{5}$

82. $\dfrac{\frac{5}{9}}{\frac{3}{4}} = \dfrac{5}{9} \cdot \dfrac{4}{3} = \dfrac{\cancel{36}^{4}(\frac{5}{\cancel{9}_{1}})}{\cancel{36}^{9}(\frac{3}{\cancel{4}_{1}})} = \dfrac{20}{27}$

83. $\dfrac{\frac{1}{2} + \frac{2}{3}}{\frac{3}{4} - \frac{1}{6}} = (\dfrac{1}{2} + \dfrac{2}{3}) \div (\dfrac{3}{4} - \dfrac{1}{6}) =$

$(\dfrac{3}{6} + \dfrac{4}{6}) \div (\dfrac{9}{12} - \dfrac{2}{12}) =$

$\dfrac{7}{6} \div \dfrac{7}{12} = \dfrac{\cancel{7}^{1}}{\cancel{6}_{1}} \cdot \dfrac{\cancel{12}^{2}}{\cancel{7}_{1}} = 2$

84. $\dfrac{3 + \frac{1}{2}}{2 - \frac{3}{4}} = (3 + \dfrac{1}{2}) \div (2 - \dfrac{3}{4}) =$

$(\dfrac{6}{2} + \dfrac{1}{2}) \div (\dfrac{8}{4} - \dfrac{3}{4}) =$

$\dfrac{7}{2} \div \dfrac{5}{4} = \dfrac{7}{\cancel{2}_{1}} \cdot \dfrac{\cancel{4}^{2}}{5} = \dfrac{14}{5} = 2\dfrac{4}{5}$

85. $5 + (1\dfrac{1}{3})(\dfrac{3}{4}) = 5 + (\dfrac{\cancel{4}^{1}}{\cancel{3}_{1}})(\dfrac{\cancel{3}^{1}}{\cancel{4}_{1}}) =$

$5 + 1 = 6$

86.
$$\frac{2}{9} + \frac{1}{3}\left(1\frac{1}{2} + 1\frac{1}{2}\right) = \frac{2}{9} + \frac{1}{3}(3) =$$

$$\frac{2}{9} + 1 = 1\frac{2}{9}$$

87.
$$\frac{12\frac{1}{3}}{6\frac{2}{3}} = \frac{\frac{37}{3}}{\frac{20}{3}} = \frac{\frac{37}{3}}{\frac{20}{3}} \cdot \frac{3}{3} =$$

$$\frac{3\left(\frac{37}{3}\right)}{3\left(\frac{20}{3}\right)} = \frac{37}{20} = 1\frac{17}{20}$$

88.
$$\frac{5 - \frac{3}{4}}{2 + \frac{3}{4}} = \frac{5 - \frac{3}{4}}{2 + \frac{3}{4}} \cdot \frac{4}{4} =$$

$$\frac{4(5) - 4\left(\frac{3}{4}\right)}{4(2) + 4\left(\frac{3}{4}\right)} = \frac{20 - 3}{8 + 3} = \frac{17}{11} = 1\frac{6}{11}$$

89.
$$\frac{\frac{3}{4} + \frac{1}{3}}{\frac{2}{3} + \frac{1}{6}} = \left(\frac{3}{4} + \frac{1}{3}\right) \div \left(\frac{2}{3} + \frac{1}{6}\right) =$$

$$\left(\frac{9}{12} + \frac{4}{12}\right) \div \left(\frac{4}{6} + \frac{1}{6}\right) = \frac{13}{12} \div \frac{5}{6} =$$

$$\frac{13}{12} \cdot \frac{6}{5} = \frac{13}{10} = 1\frac{3}{10}$$

90.
$$\frac{1 + \frac{2}{3}}{1 - \frac{2}{3}} = \frac{1 + \frac{2}{3}}{1 - \frac{2}{3}} \cdot \frac{3}{3} =$$

$$\frac{3(1) + 3\left(\frac{2}{3}\right)}{3(1) - 3\left(\frac{2}{3}\right)} = \frac{3 + 2}{3 - 2} = \frac{5}{1} = 5$$

91.
$$2y = 36$$
$$\frac{1}{2} \cdot 2y = 36^{18} \cdot \frac{1}{2}$$
$$y = 18$$

92.
$$-6x = 102$$
$$\frac{-6x}{-6} = \frac{102}{-6}$$
$$x = -17$$

93.
$$7t = -28$$
$$\frac{7t}{7} = \frac{-28}{7}$$
$$t = -4$$

94.
$$\frac{1}{4}z = 40$$
$$\frac{4}{1} \cdot \frac{1}{4}z = 40 \cdot \frac{4}{1}$$
$$z = 160$$

95.
$$-\frac{2}{3}w = 20$$
$$-\frac{2}{3} \cdot -\frac{2}{3}w = 20^{10} \cdot -\frac{3}{2}$$
$$w = -30$$

96.
$$\frac{2}{5}x = -\frac{3}{10}$$
$$\frac{5}{2} \cdot \frac{2}{5}x = -\frac{3}{10_2} \cdot \frac{5}{2}$$
$$x = -\frac{3}{4}$$

PRACTICE TEST

1a. $\dfrac{8}{3}$ the denominator is 3

b. $\dfrac{2}{9}$ the denominator is 9

c. $\dfrac{4}{17}$ the denominator is 17

d.   6   the denominator is 1

e. $\dfrac{1}{4}$ the denominator is 4

2a. $\dfrac{0}{4}$ the numerator is 0

b. $\dfrac{19}{2}$ the numerator is 19

c. $\dfrac{3}{4}$ the numerator is 3

d. $\dfrac{5}{9}$ the numerator is 5

e.   7 the numerator is 7

3a. $\dfrac{-3}{4}$ is negative

b. $\dfrac{-5}{-3}$ is positive

c. $\dfrac{7}{9}$ is positive

d. $\dfrac{1}{-6}$ is negative

4. $\dfrac{7}{8} = \dfrac{?}{64}$ $\dfrac{7}{8} \cdot \dfrac{8}{8} = \dfrac{56}{64}$

5. $\dfrac{5}{6} = \dfrac{?}{72x}$ $\dfrac{5}{6} \cdot \dfrac{12x}{12x} = \dfrac{60x}{72x}$

6. $\dfrac{56}{63} = \dfrac{2 \cdot 2 \cdot 2 \cdot 7}{3 \cdot 3 \cdot 7} = \dfrac{8}{9}$

7. $\dfrac{38}{76} = \dfrac{2 \cdot 19}{2 \cdot 2 \cdot 19} = \dfrac{1}{2}$

8. $\dfrac{8x^2y}{64xy^2} =$

$\dfrac{2 \cdot 2 \cdot 2 \cdot x \cdot x \cdot y}{2 \cdot 2 \cdot 2 \cdot 2 \cdot 2 \cdot 2 \cdot x \cdot y \cdot y} = \dfrac{x}{8y}$

9.   The recriprocal of $\dfrac{3}{5}$ is $\dfrac{5}{3}$

10.   The reciprocal of $-2\dfrac{5}{6} = -\dfrac{17}{6}$ is $-\dfrac{6}{17}$

11. $\dfrac{7}{8} + \dfrac{1}{6} = \dfrac{21}{24} + \dfrac{4}{24} = \dfrac{25}{24} = 1\dfrac{1}{24}$

12. $\left(\dfrac{17}{19}\right)\left(\dfrac{38}{51}\right) = \dfrac{\cancel{17} \cdot 2 \cdot \cancel{19}}{\cancel{19} \cdot 3 \cdot \cancel{17}} = \dfrac{2}{3}$

13. $\dfrac{1}{5} - \dfrac{4}{7} = \dfrac{7}{35} - \dfrac{20}{35} =$

$\dfrac{7 + (-20)}{35} = -\dfrac{13}{35}$

55

14. $\quad \dfrac{3}{4} + \left(-\dfrac{2}{5}\right) = \dfrac{15}{20} + \left(-\dfrac{8}{20}\right) = \dfrac{7}{20}$

15. $\quad -3\dfrac{1}{3} \div \dfrac{3}{5} = -\dfrac{10}{3} \div \dfrac{3}{5} =$

$\quad -\dfrac{10}{3} \cdot \dfrac{5}{3} = -\dfrac{50}{9} = -5\dfrac{5}{9}$

16. $\quad -4\dfrac{1}{6} \cdot \dfrac{2}{5} = -\dfrac{25}{6} \cdot \dfrac{2}{5} =$

$\quad -\dfrac{5 \cdot 5 \cdot 2}{2 \cdot 3 \cdot 5} = -\dfrac{5}{3} = -1\dfrac{2}{3}$

17. $\quad \dfrac{-1}{4} - \left(-\dfrac{11}{12}\right) = -\dfrac{1}{4} + \dfrac{11}{12} =$

$\quad \dfrac{-3}{12} + \dfrac{11}{12} = \dfrac{8}{12} = \dfrac{2}{3}$

18. $\quad \dfrac{-30x}{25} \div \dfrac{-6x}{15} = \dfrac{-30x}{25} \cdot \dfrac{15}{-6x} =$

$\quad \dfrac{-2 \cdot 3 \cdot 5 \cdot x \cdot 3 \cdot 5}{5 \cdot 5 \cdot -2 \cdot 3 \cdot x} = \dfrac{-3}{-1} = 3$

19. $\quad \left(\dfrac{2}{3} + \dfrac{3}{4}\right) + \dfrac{1}{6} = \left(\dfrac{8}{12} + \dfrac{9}{12}\right) + \dfrac{1}{6} =$

$\quad \dfrac{17}{12} + \dfrac{1}{6} = \dfrac{17}{12} + \dfrac{2}{12} = \dfrac{19}{12} = 1\dfrac{7}{12}$

20. $\quad \dfrac{\dfrac{2}{5} + \dfrac{1}{10}}{\dfrac{1}{2} - \dfrac{4}{5}} = \dfrac{\dfrac{2}{5} + \dfrac{1}{10}}{\dfrac{1}{2} - \dfrac{4}{5}} \cdot \dfrac{10}{10} =$

$\quad \dfrac{10^2\left(\dfrac{2}{5}\right) + 10\left(\dfrac{1}{10}\right)}{10_5\left(\dfrac{1}{2}\right) - 10_2\left(\dfrac{4}{5}\right)} = \dfrac{4 + 1}{5 - 8} =$

$\quad \dfrac{5}{-3} = -1\dfrac{2}{3}$

21. $\quad \dfrac{9}{8} \div \dfrac{27}{62} = \dfrac{9}{8} \cdot \dfrac{62}{27} =$

$\quad \dfrac{3 \cdot 3 \cdot 2 \cdot 31}{2 \cdot 2 \cdot 2 \cdot 3 \cdot 3 \cdot 3} = \dfrac{31}{12} = 2\dfrac{7}{12}$

22. $\quad \dfrac{3}{8} + \dfrac{5}{6} + 1\dfrac{7}{12} = \dfrac{3}{8} + \dfrac{5}{6} + \dfrac{19}{12}$

$\quad \dfrac{9}{24} + \dfrac{20}{24} + \dfrac{38}{24} = \dfrac{67}{24} = 2\dfrac{19}{24}$

23. $\quad \dfrac{4}{15} - 2\dfrac{4}{5} = \dfrac{4}{15} - \dfrac{14}{5} =$

$\quad \dfrac{4}{15} - \dfrac{42}{15} = \dfrac{4 + (-42)}{15} =$

$\quad \dfrac{-38}{15} = -2\dfrac{8}{15}$

24. $\quad -\dfrac{1}{2} \cdot \left(-\dfrac{2}{3}\right) = \dfrac{-1 \cdot -2}{2 \cdot 3} = \dfrac{1}{3}$

25. $$\frac{2}{7} + \frac{1}{4} + \frac{2}{14} =$$

$$\frac{8}{28} + \frac{7}{28} + \frac{4}{28} =$$

$\frac{19}{28}$ of the students earn an A. B, or C.

26. $$\frac{3}{8} \cdot 232 = \frac{3 \cdot 2 \cdot 2 \cdot 2 \cdot 29}{2 \cdot 2 \cdot 2} =$$

87 will earn an A.

27. $$24 \div \frac{2}{7} = \frac{24}{1} \cdot \frac{7}{2} =$$

$$\frac{2 \cdot 2 \cdot 2 \cdot 3 \cdot 7}{2} = 84 \text{ bottles}$$

28. $$1 - (\frac{1}{3} + \frac{1}{4}) = \frac{12}{12} - (\frac{4}{12} + \frac{3}{12}) =$$

$$\frac{12}{12} - (\frac{7}{12}) = \frac{5}{12} \text{ left.}$$

29. $$\frac{2}{3}x = 8$$
$$\frac{3}{2} \cdot \frac{2}{3}x = \cancel{8}^{4} \cdot \frac{3}{\cancel{2}}$$
$$x = 12$$

30. $$9 = 3x$$
$$\frac{9}{3} = \frac{3x}{3}$$
$$3 = x$$

31. $$\frac{3}{4}x = \frac{4}{3}$$
$$\frac{4}{3} \cdot \frac{3}{4}x = \frac{4}{3} \cdot \frac{4}{3}$$
$$x = \frac{16}{9} = 1\frac{7}{9}$$

32. $$\frac{9}{14} = \frac{1}{7}x$$
$$\frac{\cancel{7}}{1} \cdot \frac{9}{\cancel{14}_2} = \frac{1}{7}x \cdot \frac{7}{1}$$
$$\frac{9}{2} = x$$
$$x = 4\frac{1}{2}$$

CHAPTER 6: ALGEBRAIC EXPRESSIONS

6.1 Exponents Revisited

1. $9 = 3^2$
2. $16 = 4^2$
3. $81 = 9^2 = 3^4$
4. $32 = 2^5$

5. $\dfrac{4}{9} = (\dfrac{2}{3})^2$

6. $\dfrac{9}{4} = (\dfrac{3}{2})^2$

7. $\dfrac{16}{81} = (\dfrac{4}{9})^2$

8. $\dfrac{25}{36} = (\dfrac{5}{6})^2$

9. $\dfrac{1}{x \cdot x \cdot x \cdot x} = \dfrac{1}{x^4}$

10. $\dfrac{1}{y \cdot y \cdot y} = \dfrac{1}{y^3}$

11. $a^2b \cdot a^2b = (a^2b)^2$

12. $x^2y \cdot x^2y = (x^2y)^2$

13. $\dfrac{3pq}{z^3} \cdot \dfrac{3pq}{z^3} = (\dfrac{3pq}{z^3})^2$

14. $\dfrac{7st}{r^4} \cdot \dfrac{7st}{r^4} = (\dfrac{7st}{r^4})^2$

15. $3^0 = 1$
16. $2^0 = 1$
17. $(-15)^0 = 1$

18. $(-12)^0 = 1$
19. $(-5)^4 = 625$
20. $(-3)^6 = 729$
21. $-(12)^6 = -2{,}985{,}984$
22. $-(9)^4 = -6{,}561$
23. $(-14)^5 = -537{,}824$
24. $(-11)^3 = -1{,}331$
25. $-(8^3) = -512$
26. $-(6^5) = -7{,}776$
27. $-c^2 = -c^2$
28. $-q^4 = -q^4$
29. $(-a)^4 = a^4$
30. $(-b)^2 = b^2$
31. $(-f)^3 = -f^3$
32. $(-z)^5 = -z^5$

33. $(\dfrac{2}{3})^3 = \dfrac{8}{27}$

34. $(\dfrac{3}{4})^3 = \dfrac{27}{64}$

35. $-(\dfrac{5}{6})^2 = -\dfrac{25}{36}$

36. $-(\dfrac{7}{8})^2 = -\dfrac{49}{64}$

37. $(-\dfrac{1}{2})^2 = \dfrac{1}{4}$

38. $(-\dfrac{1}{3})^2 = \dfrac{1}{9}$

39. $6^{-3} = \dfrac{1}{6^3} = \dfrac{1}{216}$

40. $5^{-4} = \dfrac{1}{5^4} = \dfrac{1}{625}$

41. $x^{-2} = \dfrac{1}{x^2}$

42. $y^{-4} = \dfrac{1}{y^4}$

43. $(\tfrac{1}{2}q)^3 = (\tfrac{1}{2})^3 \cdot (q)^3 = \tfrac{1}{8}q^3$

44. $(\tfrac{1}{3}r)^2 = (\tfrac{1}{3})^2 \cdot r^2 = \tfrac{1}{9}r^2$

45. $(18x^2yz^3)^2 = 18^2(x^2)^2(y)^2(z^3)^2 = $

$324x^4y^2z^6$

46. $(16a^2bc^3)^2 = 16^2(a^2)^2(b)^2(c^3)^2 = $

$256a^4b^2c^6$

47. $\dfrac{(5q^2)^4}{10q^4} = \dfrac{5^4(q^2)^4}{10q^4} = $

$\dfrac{625q^8}{10q^4} = \dfrac{125q^4}{2} = 62.5q^4$

48. $\dfrac{(4p^3)^4}{16p^6} = \dfrac{4^4(p^3)^4}{4^2p^6} = $

$\dfrac{4^4 \cdot p^{12}}{4^2 \cdot p^6} = 4^2p^6 = 16p^6$

49. $(\tfrac{7}{8})^{10} = \dfrac{282,475,249}{1,073,741,824}$

50. $(\tfrac{5}{6})^8 = \dfrac{390,625}{1,679,616}$

51. $(-\tfrac{2}{3})^5 = -\dfrac{32}{243}$

52. $(-\tfrac{3}{4})^7 = -\dfrac{2,187}{16,384}$

53. $(9^5)^{-2} = 9^{-10} = \dfrac{1}{9^{10}} = \dfrac{1}{3,486,784,401}$

54. $(7^{-1})^3 = 7^{-3} = \dfrac{1}{7^3} = \dfrac{1}{343}$

55. $3^{-4} \cdot 3^{-8} = 3^{-12} = \dfrac{1}{3^{12}} = \dfrac{1}{531,441}$

56. $4^{-7} \cdot 4^{-2} = 4^{-9} = \dfrac{1}{4^9} = \dfrac{1}{262,144}$

6.2 Polynomials

1. $-x^2$ coefficient -1; 2nd degree
2. $-y^3$ coefficient -1; 3rd degree
3. $\tfrac{2}{3}x^3$ coefficient $\tfrac{2}{3}$; 3rd degree
4. $\tfrac{3}{4}y^2$ coefficient $\tfrac{3}{4}$; 2nd degree
5. 27 coefficient 27; 0 degree
6. 86 coefficient 86; 0 degree
7. $\tfrac{15}{31}$ coefficient $\tfrac{15}{31}$; 0 degree
8. $\tfrac{11}{40}$ coefficient $\tfrac{11}{40}$; 0 degree
9. $3z$ coefficient 3; 1st degree
10. $7q$ coefficient 7; 1st degree
11. $7x + 8$ is a binomial in the first degree.
12. $6y - 2$ is a binomial in the first degree.
13. $2x^2 + 15x - 3$ is a trinomial in the second degree.
14. $3y^2 - 9y + 12$ is a trinomial in the second degree.
15. $25c^3 - 18c$ is a binomial in the third degree.
16. $49r^5 + 2r$ is a binomial in the fifth degree.
17. $(3a + 17) + (14 + 2a) = $
$3a + 17 + 14 + 2a = 5a + 31$

59

18. $(21 + 5b) + (6b + 7) =$
$21 + 5b + 6b + 7 = 11b + 28$

19. $(21 + 5b) - (7 + 6b) =$
$21 + 5b - 7 - 6b = -b + 14$

20. $(3a + 17) - (2a + 14) =$
$3a + 17 - 2a - 14 = a + 3$

21. $(2x + y) - (7x + 3y) + (x + y) =$
$2x + y - 7x - 3y + x + y = -4x - y$

22. $(t + s) + (2t + 5s) - (7t + 3s) =$
$t + s + 2t + 5s - 7t - 3s = -4t + 3s$

23. Add: $t^2 + 3t + 6$
$\underline{3t^2 + 5t - 1}$
$4t^2 + 8t + 5$

24. Add: $3x^2 + 2x - 5$
$\underline{2x^2 - 6x + 7}$
$5x^2 - 4x + 2$

25. Add: $6a^4 + a^3 + 2a^2 - 5a + 7$
$\underline{a^4 \qquad\qquad + 2a + 1}$
$7a^4 + a^3 + 2a^2 - 3a + 8$

26. Add: $3c^4 + 6c^3 + c^2 - 2c + 11$
$\underline{\qquad c^3 \qquad - 5c + \ 7}$
$3c^4 + 7c^3 + c^2 - 7c + 18$

27. Subtract: $12t^2 + 6t + 7$
$\underline{\qquad t^2 + 2t + 3}$

Add: $12t^2 + 6t + 7$
$\underline{- \ t^2 - 2t - 3}$
$11t^2 + 4t + 4$

28. Subtract: $13r^2 + 4r + 3$
$\underline{\qquad r^2 + 3r + 1}$

Add: $13r^2 + 4r + 3$
$\underline{- \ r^2 - 3r - 1}$
$12r^2 + \ r + 2$

29. Subtract: $5x^4 - 2x^2 + 1$
$\underline{-2x^4 + 7x^2 - 3}$

Add: $5x^4 - 2x^2 + 1$
$\underline{2x^4 - 7x^2 + 3}$
$7x^4 - 9x^2 + 4$

30. Subtract: $4y^3 - 7y + 2$
$\underline{-3y^3 + 10y - 5}$

Add: $4y^3 - 7y + 2$
$\underline{3y^3 - 10y + 5}$
$7y^3 - 17y + 7$

31. $(4a^3 - a + 3) - (3a^3 + 6a^2 + 6a - 1) =$
$4a^3 - a + 3 - 3a^3 - 6a^2 - 6a + 1 =$
$a^3 - 6a^2 - 7a + 4$

32. $(7x^2 - 4x - 9) - (-8x^3 - 4x - 4) =$
$7x^2 - 4x - 9 + 8x^3 + 4x - 4 =$
$8x^3 + 7x^2 - 13$

33. $(3x + 6) + (4x^2 - 9) - (2x + 4) =$
$3x + 6 + 4x^2 - 9 - 2x - 4 =$
$4x^2 + x - 7$

34. $(7y + 5) + (-y^2 + 3) - (3y^2 - 6y) =$
$7y + 5 - y^2 + 3 - 3y^2 + 6y =$
$-4y^2 + 13y + 8$

35. $(3x^2) (5x^3) = 3 \cdot x^2 \cdot 5 \cdot x^2 = 15x^5$

36. $(2x^3) (4x^2) = 2 \cdot x^3 \cdot 4 \cdot x^2 = 8x^5$

37. $(-2m)^3 (6m) = (-8m^3) (6m) =$
$-8 \cdot m^3 \cdot 6 \cdot m = -48m^4$

38. $(-3n)^3 (5m) = (-27n^3) (5n) =$
$-27 \cdot n^3 \cdot 5 \cdot n = -135n^4$

39. $(3z^2) (-3z)^2 = (3z^2) (9z^2) =$
$3 \cdot z^2 \cdot 9 \cdot z^2 = 27z^4$

60

40. $(2y^2)(-2y)^2 = (2y^2)(-2y)^2 =$
$2 \cdot y^2 \cdot 4 \cdot y^2 = 8y^4$

41. $q(q + 7) = q^2 + 7q$

42. $s(s + 4) = s^2 + 4s$

43. $-3z^2(z + 1) = -3z^3 - 3z^2$

44. $-2w^2(w + 3) = -2w^3 - 6w^2$

45. $x(3x^2 + 7x + 5) = 3x^3 + 7x^2 + 5x$

46. $2x(x^2 + 3x + 1) = 2x^3 + 6x^2 + 2x$

47. $-3y(2y^2 + 7y + 1) = -6y^3 - 21y^2 - 3y$

48. $-4y(3y^2 + 2y + 10) = -12y^3 - 8y^2 - 40y$

49. $-a(ab^2 + b - 3) = -a^2b^2 - ab + 3a$

50. $-b(bc^2 + 2b - 8) = -b^2c^2 - 2b^2 + 8b$

51. $(22x^2 + 17x + 5)(-2x) =$
$-44x^3 - 34x^2 - 10x$

52. $(19y^2 + 15x + 6)(-3y) =$
$-57y^3 - 45y^2 - 18y$

53. $(x + 5)(x + 6) = x^2 + 6x + 5x + 30 =$
$x^2 + 11x + 30$

54. $(y + 3)(y + 5) = y^2 + 5y + 3y + 15 =$
$y^2 + 8y + 15$

55. $(y + 1)(2y + 7) = 2y^2 + 7y + 2y + 7 =$
$2y^2 + 9y + 7$

56. $(x + 1)(3x + 2) = 3x^2 + 2x + 3x + 2 =$
$3x^2 + 5x + 2$

57. $(3r + 5)(6r + 2) = 18r^2 + 6r + 30r + 10 =$
$18r^2 + 36r + 10$

58. $(4s + 5)(3s + 8) = 12s^2 + 32s + 15s + 40$
$= 12s^2 + 47s + 40$

59. $(7 - y)(2 + y) = 14 + 7y - 2y - y^2 =$
$14 + 5y - y^2$

60. $(4 - x)(3 + x) = 12 + 4x - 3x - x^2 =$
$12 + x - x^2$

61. $(a - 2b)(3a + b) = 3a^2 - ab - 6ab + 2b^2$
$= 3a^2 - 7ab + 2b^2$

62. $(p - 3q)(2p - q) = 2p^2 - pq - 6pq + 3q^2$
$= 2p^2 - 7pq + 3q^2$

63. $(3t + 4)(2t - 5) = 6t^2 - 15t + 8t - 20 =$
$6t^2 - 7t - 20$

64. $(2r + 9)(3r - 10) = 6r^2 - 20r + 27r - 90$
$= 6r^2 + 7r - 90$

65. $(q + 2)(q - 2) = q^2 - 2q + 2q - 4 =$
$q^2 - 4$

66. $(w + 5)(w - 5) = w^2 - 5w + 5w - 25 =$
$w^2 - 25$

67. $\left(x + \dfrac{1}{2}\right)\left(x + \dfrac{1}{4}\right) =$
$x^2 + \dfrac{1}{4}x + \dfrac{1}{2}x + \dfrac{1}{8} = x^2 + \dfrac{3}{4}x + \dfrac{1}{8}$

68. $\left(y + \dfrac{1}{5}\right)\left(y + \dfrac{1}{10}\right) =$
$y^2 + \dfrac{1}{10}y + \dfrac{1}{5}y + \dfrac{1}{50} =$
$y^2 + \dfrac{3}{10}y + \dfrac{1}{50}$

69. $3x^2 + 12x = 3x(\underline{\quad} + \underline{\quad}) =$
$3x(x + 4)$

61

70. $4y^2 + 16y = 4y\,(\underline{} + \underline{}) =$
    $4y\,(y + 4)$

71. $6z^2 + 18z + 4 = 2(3z^2 + 9z + 2)$

72. $4v^2 + 24v + 6 = 2(2v^2 + 12v + 3)$

73. $12a^2b + 2b = 2b(6a^2 + 1)$

74. $10c^2d + 5d = 5d(2c^2 + 1)$

75. $5ab^2 + 15a^2b - 35ab = 5ab(b + 3a - 7)$

76. $4g^2h - 16gh^2 + 24gh =$
    $4gh(g - 4h + 6)$

77. $-3x + 6x - 24 = 3(-x^2 + 2x - 8)$

78. $-2y^2 + 18y - 30 = 2(-y^2 + 9y - 15)$

79. $27 - 9h^2 = 9(3 - h^2)$

80. $35 - 7k^2 = 7(5 - k^2)$

6.3 Simplifying, Multiplying and Dividing Algebraic Fractions

1. $\dfrac{xy}{x^2} + \dfrac{x\cdot y}{x\cdot x} = \dfrac{y}{x}$

2. $\dfrac{cd}{d^2} = \dfrac{c\cdot d}{d\cdot d} = \dfrac{c}{d}$

3. $\dfrac{15ab^2}{10a^2b} =$

   $\dfrac{3\cdot 5\cdot a\cdot b\cdot b}{2\cdot 5\cdot a\cdot a\cdot b} = \dfrac{3b}{2a}$

4. $\dfrac{12x^2y}{20xy^2} =$

   $\dfrac{2\cdot 2\cdot 3\cdot x\cdot x\cdot y}{2\cdot 2\cdot 5\cdot x\cdot y\cdot y} = \dfrac{3x}{5y}$

5. $\dfrac{-30w}{3wx} = \dfrac{-2\cdot 3\cdot 5\cdot w}{3\cdot w\cdot x} = -\dfrac{10}{x}$

6. $\dfrac{-28q}{4qr} = \dfrac{-2\cdot 2\cdot 7\cdot q}{2\cdot 2\cdot q\cdot r} = -\dfrac{7}{r}$

7. $\dfrac{3x^2y}{12x^2y} = \dfrac{3\cdot x\cdot x\cdot y}{2\cdot 2\cdot 3\cdot x\cdot x\cdot y} = \dfrac{1}{4}$

8. $\dfrac{25cd^2}{5cd^2} = \dfrac{5\cdot 5\cdot c\cdot d\cdot d}{5\cdot c\cdot d\cdot d} = \dfrac{5}{1} = 5$

9. $\dfrac{4a}{b^2} \cdot \dfrac{3b}{2a^2} = \dfrac{2\cdot 2\cdot a\cdot 3\cdot b}{b\cdot b\cdot 2\cdot a\cdot a} = \dfrac{6}{ab}$

10. $\dfrac{6d}{f^2} \cdot \dfrac{2f}{3d^2} = \dfrac{2\cdot 3\cdot d\cdot 2\cdot f}{f\cdot f\cdot 3\cdot d\cdot d} = \dfrac{4}{df}$

11. $\dfrac{x^3}{y} \cdot \dfrac{y^5}{4x^2} =$

    $\dfrac{x\cdot x\cdot x\cdot y\cdot y\cdot y\cdot y\cdot y}{y\cdot 2\cdot 2\cdot x\cdot x} = \dfrac{xy^4}{4}$

12. $\dfrac{4x^3}{y^2} \cdot \dfrac{y^3}{x^5} =$

    $\dfrac{2\cdot 2\cdot x\cdot x\cdot x\cdot y\cdot y\cdot y}{y\cdot y\cdot x\cdot x\cdot x\cdot x\cdot x} = \dfrac{4y}{x^2}$

13. $\dfrac{3y}{2x^2} \div \dfrac{6y^2}{4x} = \dfrac{3y}{2x^2} \cdot \dfrac{4x}{6y^2} =$

$\dfrac{3 \cdot y \cdot 2 \cdot 2 \cdot x}{2 \cdot x \cdot x \cdot 2 \cdot 3 \cdot y \cdot y} = \dfrac{1}{xy}$

14. $\dfrac{4z}{3w^2} \div \dfrac{8z^2}{6w} = \dfrac{4z}{3w^2} \cdot \dfrac{6w}{8z^2} =$

$\dfrac{2 \cdot 2 \cdot z \cdot 2 \cdot 3 \cdot w}{3 \cdot w \cdot w \cdot 2 \cdot 2 \cdot 2 \cdot z \cdot z} = \dfrac{1}{wz}$

15. $\dfrac{-12t^4}{36s^3} \div \dfrac{-t}{s} = \dfrac{-12t^4}{36s^3} \div \dfrac{s}{-t} =$

$\dfrac{-2 \cdot 2 \cdot 3 \cdot t \cdot t \cdot t \cdot t \cdot 8}{3 \cdot 2 \cdot 2 \cdot 3 \cdot (-t)s \cdot s \cdot s} = \dfrac{-t^3}{-3s^2} = \dfrac{t^3}{3s^2}$

16. $\dfrac{-15s^3}{45t^4} \div \dfrac{-s}{t} = \dfrac{-15s^3}{45t^4} \cdot \dfrac{t}{-s} =$

$\dfrac{-3 \cdot 5 \cdot s \cdot s \cdot s \cdot t}{3 \cdot 3 \cdot 5 \cdot t \cdot t \cdot t \cdot t \cdot (-s)} = \dfrac{-s^2}{-3t^3} = \dfrac{s^2}{3t^3}$

17. $\dfrac{2x + 4}{9y} \cdot \dfrac{27y^2}{2} =$

$\dfrac{2(x + 2) \cdot 3 \cdot 3 \cdot 3 \cdot y \cdot y}{3 \cdot 3 \cdot y \cdot 2} =$

$\dfrac{3y(x + 2)}{1} = 3y(x + 2) = 3xy + 6y$

18. $\dfrac{3y + 6}{10x} \cdot \dfrac{25x^2}{3} =$

$\dfrac{3(y + 2) \cdot 5 \cdot 5 \cdot x \cdot x}{2 \cdot 5 \cdot x \cdot 3} =$

$\dfrac{5x(y + 2)}{2} = \dfrac{5xy + 10x}{2}$

19. $\dfrac{2ab}{7c} \cdot \dfrac{14c^2}{a^2b^2} \div \dfrac{2c}{4b} =$

$\dfrac{2ab}{7c} \cdot \dfrac{14c^2}{a^2b^2} \cdot \dfrac{4b}{2c} =$

$\dfrac{2 \cdot a \cdot b \cdot 2 \cdot 7 \cdot c \cdot c \cdot 2 \cdot 2 \cdot b}{7 \cdot c \cdot a \cdot a \cdot b \cdot b \cdot 2 \cdot c} = \dfrac{8}{a}$

20. $\dfrac{3xy}{5z} \cdot \dfrac{15z^2}{x^2y^2} \div \dfrac{6z}{2y} =$

$\dfrac{3xy}{5z} \cdot \dfrac{15z^2}{x^2y^2} \cdot \dfrac{2y}{6z} =$

$\dfrac{3 \cdot x \cdot y \cdot 3 \cdot 5 \cdot z \cdot z \cdot 2 \cdot y}{5 \cdot z \cdot x \cdot x \cdot y \cdot y \cdot 2 \cdot 3 \cdot z} = \dfrac{3}{x}$

6.4 Addition and Subtraction of Algebraic Fractions

1. $\dfrac{4c}{3} + \dfrac{c}{3} = \dfrac{5c}{3}$

2. $\dfrac{3q}{4} + \dfrac{q}{4} = \dfrac{4q}{4} = q$

3. $\dfrac{5f}{7} - \dfrac{3f}{7} = \dfrac{2f}{7}$

63

4. $\dfrac{8q}{5} - \dfrac{2q}{5} = \dfrac{6q}{5}$

5. $\dfrac{-3x}{14} + \dfrac{9x}{14} = \dfrac{6x}{14} = \dfrac{3x}{7}$

6. $\dfrac{-7y}{10} + \dfrac{17y}{10} = \dfrac{10y}{10} = y$

7. $\dfrac{5}{a} + \dfrac{2}{a} = \dfrac{7}{a}$

8. $\dfrac{6}{b} + \dfrac{4}{b} = \dfrac{10}{b}$

9. $\dfrac{3x}{10} - \dfrac{x}{10} = \dfrac{2x}{10} = \dfrac{x}{5}$

10. $\dfrac{4y}{7} - \dfrac{y}{7} = \dfrac{3y}{7}$

11. $\dfrac{x+3}{4} + \dfrac{x+2}{4} =$

$\dfrac{x+3+x+2}{4} = \dfrac{2x+5}{4}$

12. $\dfrac{y+1}{3} + \dfrac{2y+3}{3} =$

$\dfrac{y+1+2y+3}{3} = \dfrac{3y+4}{3}$

13. $\dfrac{2t+7}{5} - \dfrac{t+8}{5} = \dfrac{2t+7-(t+8)}{5} =$

$\dfrac{2t+7-t-8}{5} = \dfrac{t-1}{5}$

14. $\dfrac{6s+3}{8} - \dfrac{s+5}{8} = \dfrac{6s+3-(s+5)}{8} =$

$\dfrac{6s+3-s-5}{8} = \dfrac{5s-2}{8}$

15. $\dfrac{x}{8} + \dfrac{3}{4} = \dfrac{x}{8} + \dfrac{6}{8} = \dfrac{x+6}{8}$

16. $\dfrac{4}{5} + \dfrac{2x}{3} = \dfrac{12}{15} + \dfrac{10x}{15} = \dfrac{12+10x}{15}$

17. $\dfrac{2}{x} - \dfrac{1}{3} = \dfrac{6}{3x} - \dfrac{x}{3x} = \dfrac{6-x}{3x}$

18. $\dfrac{7}{8} - \dfrac{3}{y} = \dfrac{7y}{8y} - \dfrac{24}{8y} = \dfrac{7y-24}{8y}$

19. $\dfrac{1}{8x} + \dfrac{1}{12x} = \dfrac{3}{24x} + \dfrac{2}{24x} = \dfrac{5}{24x}$

20. $\dfrac{4}{3x} - \dfrac{3}{2x} = \dfrac{8}{6x} - \dfrac{9}{6x} = -\dfrac{1}{6x}$

21. $\dfrac{x+4}{3} + \dfrac{x+3}{2} =$

$\dfrac{2(x+4)}{6} + \dfrac{3(x+3)}{6} =$

$\dfrac{2x+8+3x+9}{6} = \dfrac{5x+17}{6}$

22. $\dfrac{2y+1}{4} + \dfrac{y+6}{5} =$

$\dfrac{5(2y+1)}{20} + \dfrac{4(y+6)}{20} =$

$\dfrac{10y+5+4y+24}{20} = \dfrac{14y+29}{20}$

64

23. $\dfrac{3}{b} - \dfrac{2}{c} = \dfrac{3c}{bc} - \dfrac{2b}{bc} = \dfrac{3c - 2b}{bc}$

24. $\dfrac{12}{x} - \dfrac{13}{y} = \dfrac{12y}{xy} - \dfrac{13x}{xy} =$

$\dfrac{12y - 13x}{xy}$

25. $\dfrac{2x}{y} + \dfrac{3y}{x} = \dfrac{2x^2}{xy} + \dfrac{3y^2}{xy} =$

$\dfrac{2x^2 + 3y^2}{xy}$

26. $\dfrac{5v}{w} + \dfrac{7w}{v} = \dfrac{5v^2}{vw} + \dfrac{7w^2}{vw} =$

$\dfrac{5v^2 + 7w^2}{vw}$

27. $\dfrac{1}{2xy} + \dfrac{3y}{x^2} + \dfrac{5x}{y^2} =$

$\dfrac{xy}{2x^2y^2} + \dfrac{6y^3}{2x^2y^2} + \dfrac{10x^3}{2x^2y^2} =$

$\dfrac{xy + 6y^3 + 10x^3}{2x^2y^2}$

28. $\dfrac{2a}{b^2} + \dfrac{3b}{a^2} + \dfrac{1}{3ab} =$

$\dfrac{6a^3}{3a^2b^2} + \dfrac{9b^3}{3a^2b^2} + \dfrac{ab}{3a^2b^2} =$

$\dfrac{6a^3 + 9b^3 + ab}{3a^2b^2}$

29. $\dfrac{10s}{11t} - \dfrac{1}{11} + \dfrac{3t}{s} =$

$\dfrac{10s^2}{11st} - \dfrac{st}{11st} + \dfrac{33t^2}{11st} =$

$\dfrac{10s^2 - st + 33t^2}{11st}$

30. $\dfrac{1}{3p} - \dfrac{2p}{q} + \dfrac{q}{3} =$

$\dfrac{q}{3pq} - \dfrac{6p^2}{3pq} + \dfrac{pq^2}{3pq} =$

$\dfrac{q - 6p^2 + pq^2}{3pq}$

6.5 Evaluating Algebraic Expressions

1. $3 + t$, when $t = 2$

t	$3 + t$	Answer
2	$3 + 2$	5

2. $4 + s$, when $s = 3$

s	$4 + s$	Answer
3	$4 + 3$	7

3. $4 + s$, when $s = -3$

s	$4 + 3$	Answer
-3	$4 + (-3)$	1

65

4. $3 + t$, when $t = -2$

t	$3 + t$	Answer
-2	$3 + (-2)$	1

5. $13r - 1$, when $r = 5$

r	$13r - 1$	Answer
5	$13(5) - 1 = 65 - 1$	64

6. $12q - 3$, when $q = 7$

q	$12 - 3$	Answer
7	$12(7) - 3 = 84 - 3$	81

7. $3z + 7$, when $z = -22$

z	$3z + 7$	Answer
-22	$3(-22) + 7 = -66 + 7$	-59

8. $2w + 8$, when $w = -18$

w	$2w + 8$	Answer
-18	$2(-18) + 8 = -36 + 8$	-28

9. $10x + 5y$, when $x = 2$, $y = -1$

x; y	$10x + 5y$	Answer
2; -1	$10(2) + 5(-1) = 20 + (-5)$	15

10. $5x + 10y$, when $x = 2$, $y = -1$

x; y	$5x + 10y$	Answer
2; -1	$5(2) + 10(-1) = 10 + -10$	0

11. $2x^2 + 6x + 6$, when $x = 2$

x	$2x^2 + 6x + 5$	Answer
3	$2(3)^2 + 6(3) + 5 =$ $18 + 18 + 5$	41

12. $3x^2 + 5x + 6$, when $x = 2$

x	$3x^2 + 5x + 6$	Answer
2	$3(2)^2 + 5(2) + 6 =$ $12 + 10 + 6$	28

13. $3x^2 + 5x + 6$, when $x = 1$

x	$3x^2 + 5x + 6$	Answer
-1	$3(-1)^2 + 5(-1) + 6 =$ $3 + (-5) + 6$	4

14. $2x^2 + 6x + 5$, when $x = -1$

x	$2x^2 + 6x + 5$	Answer
-1	$2(-1)^2 + 6(-1) + 5 =$ $2 + (-6) + 5$	1

15. $\dfrac{a + b}{2}$, when a = 14, b = 12

a, b	$\dfrac{a + b}{2}$	Answer
14,12	$\dfrac{14 + 12}{2} = \dfrac{26}{2}$	13

16. $\dfrac{c + d}{3}$, when c = 13, d = 14

c, d	$\dfrac{c + d}{3}$	Answer
13,14	$\dfrac{13 + 4}{3} = \dfrac{27}{3}$	9

17. $3(x + 7) - 5$, when x = 13

x	$3(x + 7) - 5$	Answer
13	$3(13 + 7) - 5 = 60 - 5$	55

18. $4(y + 6) - 8$, when y = 14

y	$4(y + 6) - 8$	Answer
14	$4(14 + 6) - 8 = 80 - 8$	72

19. $\dfrac{3c - 4f}{2e}$, when $c = -1, f = 3, e = 4$

c, f, e	$\dfrac{3c - 4f}{2e}$	Answer
-1, 3, 4	$\dfrac{3(-1) - 4(3)}{2(4)} =$ $\dfrac{-3 - 12}{8}$	$\dfrac{-15}{8}$

20. $\dfrac{5q - 3s}{4t}$, when $q = 2, s = -1, t = 3$

q, s, t	$\dfrac{5q - 3s}{4t}$	Answer
2, -1, 3	$\dfrac{5(2) - 3(-1)}{4(3)} =$ $\dfrac{10 + 3}{12}$	$\dfrac{13}{12}$

21. $\dfrac{1}{2}x^3 + \dfrac{2}{3}x^2 - 15x + 6$, when $x = \dfrac{1}{3}$

$= \dfrac{1}{2}(\dfrac{1}{3})^3 + \dfrac{2}{3}(\dfrac{1}{3})^2 - 15(\dfrac{1}{3}) + 6$

$= \dfrac{1}{54} + \dfrac{2}{27} - 5 + 6$

$= \dfrac{1}{54} + \dfrac{4}{54} - 5 + 6 = 1\dfrac{5}{54}$

22. $\frac{1}{3}y^3 + \frac{3}{4}y^2 - 12y + 7$, when $y = \frac{1}{2}$

$= \frac{1}{3}(\frac{1}{2})^3 + \frac{3}{4}(\frac{1}{2})^2 - 12(\frac{1}{2}) + 7$

$= \frac{1}{24} + \frac{3}{16} - 6 + 7$

$= \frac{2}{48} + \frac{9}{48} - 6 + 7 = 1\frac{11}{48}$

23. $\frac{5}{6}a^2 + \frac{7}{8}a + 11$, when $a = -\frac{2}{3}$

$= \frac{5}{6}(-\frac{2}{3})^2 + \frac{7}{8}(-\frac{2}{3}) + 11$

$= \frac{5}{27} + \frac{7}{12} + 11$

$= \frac{20}{108} + \frac{63}{108} + 11 = 11\frac{83}{108}$

24. $\frac{2}{3}b^2 + \frac{8}{9}b + 15$, when $b = \frac{1}{2}$

$= \frac{2}{3}(\frac{1}{4})^2 + \frac{8}{9}(\frac{1}{4}) + 15$

$= \frac{1}{24} + \frac{2}{9} + 15$

$= \frac{3}{72} + \frac{16}{72} + 15 = 15\frac{19}{72}$

25. $1\frac{1}{10}t^2 - 3\frac{2}{3}t + 5\frac{3}{4}$, when $t = 2\frac{1}{3}$

$= \frac{11}{10}t^2 - \frac{11}{3}t + \frac{23}{4}$, when $t = \frac{7}{3}$

$= \frac{11}{10}(\frac{7}{3})^2 - \frac{11}{3}(\frac{7}{3}) + \frac{23}{4}$

$= \frac{539}{90} - \frac{77}{9} + \frac{23}{4} = 3\frac{11}{60}$

26. $7\frac{1}{5}w^2 - 1\frac{1}{2}w + 3\frac{9}{10}$, when $w = -4\frac{1}{4}$

$= 7\frac{1}{5}(-4\frac{1}{4})^2 - 1\frac{1}{2}(-4\frac{1}{4}) + 3\frac{9}{10}$

$= 130\frac{1}{20} - (-6\frac{3}{8}) + 3\frac{9}{10} = 140\frac{13}{40}$

27. $\frac{a-b}{c}$, when $a = \frac{1}{2}$, $b = \frac{1}{3}$, $c = \frac{1}{4}$

$= \frac{\frac{1}{2} - \frac{1}{3}}{\frac{1}{4}} = \frac{\frac{1}{6}}{\frac{1}{4}} = \frac{2}{3}$

28. $\frac{w+y}{z}$, when $w = \frac{1}{5}$, $y = -\frac{1}{6}$, $z = \frac{1}{2}$

$= \frac{\frac{1}{5} + (-\frac{1}{6})}{\frac{1}{2}} = \frac{\frac{1}{30}}{\frac{1}{2}} = \frac{1}{15}$

29. $\dfrac{jk - k^2}{3j}$, when $j = 27$, $k = 31$

$= \dfrac{(27)(31) - 31^2}{3(27)} = \dfrac{837 - 961}{81}$

$= -\dfrac{124}{81} = -1\dfrac{43}{81}$

30. $\dfrac{xy - x^2}{4y}$, when $x = 13$, $y = 25$

$= \dfrac{(13)(25) - (13)^2}{4(25)} = \dfrac{325 - 169}{100}$

$= \dfrac{156}{100} = \dfrac{39}{25} = 1\dfrac{14}{25}$

31. $P = 2L + 2W$, find P
when $L = 18$ and $W = 6$

$P = 2(18) + 2(6)$
$P = 36 + 12$
$P = 48$

32. $A = L \cdot W$, find A
when $L = 20$ and $W = 11$

$A = 20 \cdot 11$
$A = 220$

33. $A = \pi r^2$, find A when $r = 3$ ($\pi \approx 3.14$)

$A = 3.14(3)^2$
$A = 28.26$

34. $C = d\pi$, find C when $d = 5$ ($\pi = 3.14$)

$C = 5(3.14)$
$C = 15.7$

35. $A = \dfrac{1}{2}bh$, find A when $b = 4$, $h = 1$

$A = \dfrac{1}{2}(4)(1)$

$A = 2$

36. $V = \dfrac{4}{3}\pi r^3$, find V when $r = 6$

$V = \dfrac{4}{3}(3.14)(6)^3$

$V = 904.32$

37. $C = \dfrac{5}{9}(f - 32)$, find C when $F = 54$

$C = \dfrac{5}{9}(54 - 32)$

$C = \dfrac{5}{9}(22) = 12\dfrac{2}{9}° = 12.2°$ celsius

38. $C = \dfrac{5}{9}(f - 32)$, find C when $F = 88$

$C = \dfrac{5}{9}(88 - 32)$

$C = \dfrac{5}{9}(56)$

$C = 31\dfrac{1}{9}° = 31.1°$ celsius

REVIEW

1.   See Section 6.1
2.   See Section 6.1
3.   See Section 6.1
4.   See Section 6.1
5.   See Section 6.2

6.   See Section 6.1
7.   See Section 6.1
8.   See Section 6.1
9.   See Section 6.1
10. See Section 6.2
11. See Section 6.2
12. See Section 6.2
13. See Section 6.4
14. See Section 6.4
15. See Section 6.5

16. $\dfrac{125}{32}$ in exponential form is $\dfrac{5^3}{2^5}$

17. $\dfrac{x \cdot x \cdot x}{3 \cdot 3 \cdot 3}$ in exponential form is

$$\dfrac{x^3}{3^3} = \left(\dfrac{x}{3}\right)^3$$

18. $\left(-\dfrac{5}{6}\right)^3 = -\dfrac{125}{216}$

19. $3x^2 + 7x + 1$ is a trinomial in the 2nd degree.

20. $\dfrac{7}{8}$ is a monomial in the 0 degree.

21. $2x - 3$ is a binomial in the 1st degree.

22. $\dfrac{x^3}{3}$ is a monomial in the 3rd degree.

23. $(17y + 5) + (2y - 18) =$
$17y + 5 + 2y - 18 = 19y - 13$

24. $(3x + 11) - (5x - 7) =$
$3x + 11 - 5x + 7 = -2x + 18$

25. Add: $4x^2 + 12x - 1$
$\underline{3x^2 - 5y + 5}$
$7x^2 - 15x + 4$

26. Subtract: $16y^2 - 12y + 23$
$\underline{11y^2 - 5y + 38}$
Add: $16y^2 - 12y + 23$
$\underline{-11y^2 + 5y - 38}$
$5y^2 - 7y - 15$

27. $-3y(5y^2 - 7y + 13) =$
$-15y^3 + 21y^2 - 39y$

28. $(3z^2 + z + 2)(5z) = 15z^3 + 5z^2 + 10z$

29. $(-x)(2x^2 + x - 1) = -2x^3 - x^2 + x$

30. $(z + 5)(z + 6) = z^2 + 6z + 5z + 30 =$
$z^2 + 11z + 30$

31. $(3x + 1)(9x - 2) = 27x^2 - 6x + 9x - 2 =$
$27x^2 + 3x - 2$

32. $(q - 2)(q + 2) = q^2 + 2q - 2q - 4 =$
$q^2 - 4$

33. $(2x - 3y)(3x - 2y) =$
$6x^2 - 4xy - 9xy + 6y =$
$6x^2 - 13xy + 6y^2$

34. $6w + 12x - 4y = 2(3w + 6x - 2y)$
②$3 \cdot w$ ②$2 \cdot 3 \cdot x$ ②$2 \cdot y$

35. $12y + 3y^2 = 3y(4 + y)$
$2 \cdot 2 ③Ⓨ ③ \cdot y Ⓨ$

36. $-15x^2y + 25xy^2 - 5xy =$
$5xy(-3x + 5y - 1)$
$3 ⑤Ⓧ xⓎ ⑤ 5 ⓍⓎ y ⑤ⓍⓎ$

70

37. $\dfrac{-24x}{30x^2y} = \dfrac{-2 \cdot 2 \cdot 2 \cdot 3 \cdot 4}{2 \cdot 3 \cdot 5 \cdot x \cdot x \cdot y} =$

$\dfrac{-4}{5xy} = -\dfrac{4}{5xy}$

38. $\dfrac{520gh}{-5g^2} = \dfrac{2 \cdot 2 \cdot 2 \cdot 5 \cdot 13 \cdot g \cdot h}{-5 \cdot g \cdot g} =$

$\dfrac{104h}{-g} = -\dfrac{104h}{g}$

39. $\dfrac{2}{x} \cdot \dfrac{y}{x} = \dfrac{2y}{x^2}$

40. $\dfrac{3h^2}{4j} \cdot \dfrac{12j^2}{6h} = \dfrac{3 \cdot h \cdot h \cdot 2 \cdot 2 \cdot 3 \cdot j \cdot j}{2 \cdot 2 \cdot j \cdot 2 \cdot 3 \cdot h} =$

$\dfrac{3hj}{2}$

41. $\dfrac{6s}{3r} \cdot \dfrac{9r^2}{3s^2} =$

$\dfrac{2 \cdot 3 \cdot 5 \cdot 3 \cdot 3 \cdot r \cdot r}{3 \cdot r \cdot 3 \cdot s \cdot s} = \dfrac{6r}{s}$

42. $\dfrac{6}{r^2} \div \dfrac{12}{r} = \dfrac{6}{r^2} \cdot \dfrac{r}{12} =$

$\dfrac{2 \cdot 3 \cdot r}{r \cdot r \cdot 2 \cdot 2 \cdot 3} = \dfrac{1}{2r}$

43. $\dfrac{35c^2d}{5c} \div \dfrac{7cd}{10d^2} = \dfrac{35c^2d}{5c} \cdot \dfrac{10d^2}{7cd} =$

$\dfrac{5 \cdot 7 \cdot c \cdot c \cdot d \cdot 2 \cdot 5 \cdot d \cdot d}{5 \cdot c \cdot 7 \cdot c \cdot d} =$

$\dfrac{10d^2}{1} = 10d^2$

44. $\dfrac{3}{a} + \dfrac{2}{a} = \dfrac{5}{a}$

45. $\dfrac{6}{xy} - \dfrac{4}{xy} = \dfrac{2}{xy}$

46. $\dfrac{6x}{2c} + \dfrac{10x}{2c} = \dfrac{16x}{2c} = \dfrac{8x}{c}$

47. $\dfrac{11y}{5x} - \dfrac{y}{5x} = \dfrac{10y}{5x} = \dfrac{2y}{x}$

48. $\dfrac{7a+14}{5b} - \dfrac{2a+4}{5b} =$

$\dfrac{7a+14-2a-4}{5b} = \dfrac{5a+10}{5b} =$

$\dfrac{5(a+2)}{5(b)} = \dfrac{a+2}{b}$

49. $\dfrac{a}{3} + \dfrac{a}{2} = \dfrac{2a}{6} + \dfrac{3a}{6} = \dfrac{5a}{6}$

50. $\dfrac{3b}{10} - \dfrac{b}{5} = \dfrac{3b}{10} - \dfrac{2b}{10} = \dfrac{b}{10}$

51. $\dfrac{x}{4} + \dfrac{x^2}{6} = \dfrac{3x}{12} + \dfrac{2x^2}{12} = \dfrac{3x+2x^2}{12}$

52. $\dfrac{13r}{2s} - \dfrac{r}{3s} = \dfrac{39r}{6s} - \dfrac{2r}{6s} = \dfrac{37r}{6s}$

71

53. $\dfrac{2}{y} + \dfrac{1-y}{y} = \dfrac{2+1-y}{y} = \dfrac{3-y}{y}$

54. $\dfrac{3}{w} + \dfrac{5}{w} - \dfrac{2}{w} = \dfrac{6}{w}$

55. $\dfrac{17}{x^2} + \dfrac{8}{y} + \dfrac{10}{x} =$

$\dfrac{17y}{x^2y} + \dfrac{8x^2}{x^2y} + \dfrac{10xy}{x^2y} =$

$\dfrac{17y + 8x^2 + 10xy}{x^2y}$

56. $2x + 1$, for $x = 3$
$2(3) + 1 = 6 + 1 = 7$

$2x + 1$, for $x = -1$
$2(-1) + 1 = -2 + 1 = -1$

57. $3a + 5b$, for $a = 2$ and $b = 7$
$3(2) + 5(7) = 6 + 35 = 41$

$3a + 5b$, for $a = -3$ and $b = 5$
$3(-3) + 5(5) = -9 + 25 = 16$

58. $6x^4 + 5x^2 - 3x + 11$, for $x = 1$
$6(1)^4 + 5(1)^2 - 3(1) + 11 =$
$6 + 5 - 3 + 11 = 19$

$6x^4 + 5x^2 - 3x + 11$, for $x = -4$
$6(-4)^4 + 5(-4)^2 - 3(-4) + 11 =$
$1536 + 80 + 12 + 11 = 1639$

59. $\dfrac{1}{2}y^3 + \dfrac{2}{3}y^2 - \dfrac{1}{5}y + \dfrac{3}{8}$, for $y = 30$

$\dfrac{1}{2}(30)^3 + \dfrac{2}{3}(30)^2 - \dfrac{1}{5}(30) + \dfrac{3}{8} =$

$13{,}500 + 600 - 6 + \dfrac{3}{8} = 14{,}904\dfrac{3}{8}$

60. $\dfrac{3w + 4x - 5y}{2z}$,

for $w = 5$, $x = -4$, $y = 3$, $z = -2 =$

$\dfrac{3(5) + 4(-4) - 5(3)}{2(-2)} =$

$\dfrac{15 + (-16) - 15}{-4} = \dfrac{-16}{-4} = 4$

PRACTICE TEST

1. $(10x + 2y) + (7x - 3y) =$
$10x + 2y + 7x - 3y = 17x - y$

2. $(a + 5b)(2a - b) = 2a^2 - ab + 10ab - 5b^2$
$= 2a^2 + 9ab - 5b^2$

3. $(2x + 3y)(3x + 2y) =$
$6x^2 + 4xy + 9xy + 6y^2 = 6x^2 + 13xy + 6y^2$

4. $(c + 4d)(c - 4d) =$
$c^2 - 4cd + 4cd - 16d^2 = c^2 - 16d^2$

5. $x^2(5x^2 + 21x - 11) = 5x^4 + 21x^3 - 11x^2$

6. $(-2)(y^2 - 10y + 5) = -2y^2 + 20y - 10$

7. $(ab)(a^2 + 3ab + b^2) = a^3b + 3a^2b^2 + 3ab^3$

8. (17v − 2w) − (23v + 5w)
    17v − 2w − 23v − 5w = −6v − 7w

9. (3a + 54b − 6c) − (31a − 42b + 19c) =
    3a + 54b − 6c − 31a + 42b − 19c =
    −28a + 96b −25c

10. (5x² + 2x + 6) − (x² − 12x + 18) =
    5x² + 2x + 6 − x² + 12x − 18) =
    4x² + 14x − 12

11. $\dfrac{15}{x} + \dfrac{12}{y} = \dfrac{15y}{xy} + \dfrac{12x}{xy} =$

    $\dfrac{15y + 12x}{xy}$

12. $\dfrac{3z}{2xy} - \dfrac{1}{xy} = \dfrac{3z}{2xy} - \dfrac{2}{2xy} =$

    $\dfrac{3z - 2}{2xy}$

13. $\dfrac{2a + b}{5} - \dfrac{a - b}{2} =$

    $\dfrac{2(2a + b)}{10} - \dfrac{5(a - b)}{10} =$

    $\dfrac{4a + 2b - 5a + 5b}{10} = \dfrac{-a + 7b}{10}$

14. −2(25x² − 6x + 33) = −50x² + 12x − 66

15. $\dfrac{x}{y} \div \dfrac{y}{x} = \dfrac{x}{y} \cdot \dfrac{x}{y} = \dfrac{x^2}{y^2}$

16. $\dfrac{3a^2}{b} \div \dfrac{b^2}{8a} = \dfrac{3a^2}{b} \cdot \dfrac{8a}{b^2} = \dfrac{24a^3}{b^3}$

17. $\dfrac{-6k}{m^3} \div \dfrac{-2k}{m} = \dfrac{-6k}{m^3} \cdot \dfrac{m}{-2k} =$

    $\dfrac{2 \cdot 3 \cdot k \cdot m}{m \cdot m \cdot m \cdot -2 \cdot k} = \dfrac{-3}{-m^2} = \dfrac{3}{m^3}$

18. $\dfrac{-32r^2s}{18r^3s^5} =$

    $\dfrac{2 \cdot 2 \cdot 2 \cdot 2 \cdot 2 \cdot r \cdot r \cdot s}{2 \cdot 3 \cdot 3 \cdot r \cdot r \cdot r \cdot s \cdot s \cdot s \cdot s \cdot s} = \dfrac{-16}{9rs^4}$

19. $\dfrac{3a}{2b} + \dfrac{5a}{2b} = \dfrac{8a}{2b} = \dfrac{4a}{b}$

20. $(4q^3)^2 = 16q^6$

21. $(r^4)^{-1} = \dfrac{1}{(r^4)^1} = \dfrac{1}{r^4}$

22. $(\dfrac{2g}{3h})^3 = \dfrac{8g^3}{27h^3}$

23. $x^2(x^{15}) = x^{17}$

24. $-(5x)^2 = -(5x)(5x) = -25x^2$

25. $(72y)^0 = 1$

26. $(4ab^3)^{-2} = \dfrac{1}{(4ab^3)^2} = \dfrac{1}{16a^2b^6}$

27. $(-3q)^2 = 9q^2$

28. $5x^{-7} = \dfrac{5}{x^7}$

73

29. $16 - 5y$, for $y = 5$
$16 - 5(5) = 16 - 25 = -9$

30. $7b + 9c$, for $b = 8$ and $c = -6$
$7(8) + 9(-6) = 56 + (-54) = 2$

31. $4x^2 - 10x + 8$, for $x = 3$
$4(3)^2 - 10(3) + 8 = 36 - 30 + 8 = 14$

32. $3y^2 - 7y + 21$, for $y = -5$
$3(-5)^2 - 7(-5) + 21 =$
$75 + 35 + 21 = 131$

33. $2a^2 + 3ab + b^2$, for $a = -1$ and $b = 4$
$2(-1)^2 + 3(-1)(4) + (4)^2 =$
$2 - 12 + 16 = 6$

34. $10^{-1} = \dfrac{1}{10^1} = \dfrac{1}{10}$

35. $(\dfrac{3}{4})^3 = \dfrac{3^3}{4^3} = \dfrac{27}{64}$

36. $2^{-2} \cdot 3^{-3} = \dfrac{1}{2^2} \cdot \dfrac{1}{3^3} =$

           $\dfrac{1}{4} \cdot \dfrac{1}{27} = \dfrac{1}{108}$

37. $-\dfrac{5}{6}$ is a monomial to the 0 degree.

38. $2t + 3$ is a binomial to the 1st degree.

39. $17x^4 - 5x^2 + 81$ is a trinomial to the 4th degree.

40. $-5x^2 + 27$ is a binomial to the 2nd degree.

CUMULATIVE REVIEW: CHAPTERS 4, 5, AND 6

Part I

1. The letter "x" when used in mathematics is an example of a <u>variable</u>.

2. <u>The Identity Property of Multiplication</u> states that, for all real numbers "a," $a \cdot 1 = a$.

3. A polynomial with three terms is called a <u>trinomial</u>.

4. $6x + 9 \times (-9) = 7 + (-9)$ is an example of the <u>Addition Property of Equality</u>.

5. $14x^4 - 3x^2 + 7x + 18$ is an example of a <u>polynomial</u>.

6. <u>The Property of Additive Inverses</u> states that, for all $a \in$ Reals, $a + (-a) = 0$.

7. $(14 + 7)$ and $(-8b + 3)$ are examples of <u>binomials</u>.

8. <u>The Distributive Property</u> states that, for all $a, b, c, \in$ Reals, $a(b + c) = ab + ac$ and $ab + ac = a(b + c)$.

9. $\frac{1}{7} \cdot 7x = 14 \cdot \frac{1}{7}$ is an example of the <u>Multiplication Property of Equality</u>.

10. <u>The Property of Reciprocals</u> states that, for all real numbers, $a(a \neq 0)$, $a \cdot \frac{1}{a} = 1$.

Part II

11. Evaluate: $2y^2 - y + 3$, when $y = -3$

$2(y)(y) - (y) + 3 = 2(-3)(-3) - (-3) + 3$
$= 18 + 3 + 3 = 24$

12. Evaluate: $2u - v^3$, when $u = 33$ and $v = 2$

$2u - \dfrac{1}{v^3} = 2u - \dfrac{1}{(v)(v)(v)} =$

$2(33) - \dfrac{1}{(2)(2)(2)} = 66 - \dfrac{1}{8} = 65\dfrac{7}{8}$

13. $\dfrac{25}{7} \cdot 3\dfrac{4}{7}$

14. $-(-7) = 7$

15. $\dfrac{0}{4} = 0$

16. $\dfrac{4}{0} =$ **undefined**

17. $(-6)(-4) = 24$

18. $7(-3) = -21$

19. The reciprocal of $-6\dfrac{2}{7}$ is $-\dfrac{7}{44}$

20. $6 - (-5) = 6 + 5 = 11$

21. $-22 + (-11) = -33$

22. $\dfrac{17}{34} = \dfrac{1}{2}$

74a

23. $\dfrac{9^8}{9^3} = 9^5$

24. $-\dfrac{3}{7} + \dfrac{5}{4} = -\dfrac{12}{28} + \dfrac{35}{28} = -\dfrac{23}{28}$

25. $-2\dfrac{3}{4} + 5\dfrac{4}{5}$

$5\dfrac{4}{5} = 5\dfrac{16}{20}$

$-2\dfrac{3}{4} = 2\dfrac{15}{20} = 3\dfrac{1}{20}$

26. $\dfrac{\dfrac{1}{3} + \dfrac{1}{4}}{\dfrac{5}{6} + \dfrac{1}{12}} = \dfrac{12^4(\frac{1}{3}) + 12^3(\frac{1}{4})}{12_2(\frac{5}{6}) + 12(\frac{1}{12})} =$

$\dfrac{4 + 3}{10 + 1} = \dfrac{7}{11}$

27. $28 + 7 - 6 - (-8) =$
$28 + 7 + (-6) + 8 = 37$

28. $\{21 - [5 - (7 - 2 \div 2)]\} \div (3 + 16 \div 2) =$
$\{21 - [5 - (7 - 1)]\} \div (3 + 8) =$
$\{21 - [5 - 6]\} \div (11) =$
$\{21 - [-1]\} \div (11) = \{21 + 1\} \div 11 =$
$22 \div 11 = 2$

29. $(-2m^5 - 7m^3 - 5) + (-5m^5 - 2m^3) =$
$-2m^5 - 7m^3 - 5 + 5m^5 - 2m^3 =$
$3m^5 - 9m^3 - 5$

30. $(8x^3)(4xy^2) = 32x^4y^2$

31. $(-3x^3y)^2 = 9x^6y^2$

32. $(x + 2)(x + 1) =$
$x^2 + x + 2x + 2 = x^2 + 3x + 2$

33. $(4a - 2)(a + 1) =$
$4a^2 + 4a - 2a - 2 = 4a^2 + 2a - 2$

34. $2x^3 + 3x = x(2x^2 + 3)$
$2 \cdot x \cdot x \,\textcircled{x}\, 3 \,\textcircled{x}$

35. $(2x)^0 = 1$

36. $\dfrac{x^6}{x^5} = x^1 = x$

37. $\dfrac{k^{10}m^{14}}{k^8} = k^2m^{14}$

38. $\dfrac{3x^2}{18} \cdot \dfrac{6}{x^3} = \dfrac{1}{x}$

39. $\dfrac{3}{4} \div \dfrac{9}{x} = \dfrac{3}{4} \cdot \dfrac{x}{9} = \dfrac{x}{12}$

40. $\dfrac{7}{12n^2} \div \dfrac{35}{n^3} = \dfrac{7}{12n^2} \cdot \dfrac{n^3}{35} = \dfrac{n}{60}$

41. $\dfrac{181y}{11} + \dfrac{27y}{11} = \dfrac{208y}{11}$

42. $\dfrac{3a}{5} + \dfrac{a}{7} = \dfrac{21a}{35} + \dfrac{5a}{35} = \dfrac{26a}{35}$

43. $\dfrac{4}{x^2} + \dfrac{6}{x^3} = \dfrac{4x}{x^3} + \dfrac{6}{x^3} = \dfrac{4x + 6}{x^3}$

44. $\dfrac{1}{a} + \dfrac{1}{b} = \dfrac{b}{ab} + \dfrac{a}{ab} = \dfrac{b + a}{ab}$

45. $23 + a = \dfrac{36}{6}$

$23 + a = 6$
$-23 \quad\quad = -23$
$\quad\quad a = -17$

74b

46.
$$x + 8 - 3 = 5 + x$$
$$x + 5 = 5 + x$$
$$-5 = -5$$
$$x = x$$

any real number

47.
$$-\frac{1}{5}y = 3$$
$$-\frac{5}{1} \cdot -\frac{1}{5}y = 3 \cdot -\frac{5}{1}$$
$$y = -15$$

48.
$$7x = 56$$
$$\frac{1}{7} \cdot 7x = 56 \cdot \frac{1}{7}$$
$$x = 8$$

74c

CUMULATIVE REVIEW: CHAPTERS 4, 5, AND 6

1. The letter "x" when used in mathematics is an example of a <u>variable</u>.

2. <u>The Identity Property of Multiplication</u> states that, for all real numbers "a," $a \cdot 1 = a$.

3. A polynomial with three terms is called a <u>trinomial</u>.

4. $6x + 9$ X $(-9) = 7 + (-9)$ is an example of the <u>Addition Property of Equality</u>.

5. $14x^4 - 3x^2 + 7x + 18$ is an example of a <u>polynomial</u>.

6. <u>The Property of Additive Inverses</u> states that, for all $a \in$ Reals, $a + (-a) = 0$.

7. $(14 + 7)$ and $(-8b + 3)$ are examples of <u>binomials</u>.

8. <u>The Distributive Property</u> states that, for all a, b, c, $\in$ Reals, $a(b + c) = ab + ac$ and $ab + ac = a(b + c)$.

9. $\frac{1}{7} \cdot 7x = 14 \cdot \frac{1}{7}$ is an example of the <u>Multiplication Property of Equality</u>.

10. <u>The Property of Reciprocals</u> states that, for all real numbers, $a(a \neq 0)$, $a \cdot \dfrac{1}{a} = 1$.

11. Evaluate: $2y^2 - y + 3$, when $y = -3$

$2(y)(y) - (y) + 3 = 2(-3)(-3) - (-3) + 3$
$= 18 + 3 + 3 = 24$

12. Evaluate: $2u - v^{-3}$, when u = 33 and v = 2

$2u - \dfrac{1}{v^3} = 2u - \dfrac{1}{(v)(v)(v)} =$

$2(33) - \dfrac{1}{(2)(2)(2)} = 66 - \dfrac{1}{8} = 65\dfrac{7}{8}$

13. $\dfrac{25}{7} \cdot 3\dfrac{4}{7}$

14. $-(-7) = 7$

15. $\dfrac{0}{4} = 0$

16. $\dfrac{4}{0} =$ **undefined**

17. $(-6)(-4) = 24$

18. $7(-3) = -21$

19. The reciprocal of $-6\dfrac{2}{7}$ is $-\dfrac{7}{44}$

20. $6 - (-5) = 6 + 5 = 11$

21. $-22 + (-11) = -33$

22. $\dfrac{17}{34} = \dfrac{1}{2}$

23. $\dfrac{9^8}{9^3} = 9^5$

24. $-\dfrac{3}{7} + \dfrac{5}{4} = -\dfrac{12}{28} + \dfrac{35}{28} = -\dfrac{23}{28}$

25. $-2\frac{3}{4} + 5\frac{4}{5}$

$5\frac{4}{5} = 5\frac{16}{20}$

$-2\frac{3}{4} = 2\frac{15}{20} = 3\frac{1}{20}$

26. $\dfrac{\frac{1}{3} + \frac{1}{4}}{\frac{5}{6} + \frac{1}{12}} = \dfrac{\cancel{12}^{4}(\frac{1}{\cancel{3}}) + \cancel{12}^{3}(\frac{1}{\cancel{4}})}{\cancel{12}_{2}(\frac{5}{\cancel{6}}) + \cancel{12}(\frac{1}{\cancel{12}})} =$

$\dfrac{4 + 3}{10 + 1} = \dfrac{7}{11}$

27. $28 + 7 - 6 - (-8) =$
$28 + 7 + (-6) + 8 = 37$

28. $\{21 - [5 - (7 - 2 \div 2)]\} \div (3 + 16 \div 2) =$
$\{21 - [5 - (7 - 1)]\} \div (3 + 8) =$
$\{21 - [5 - 6]\} \div (11) =$
$\{21 - [-1]\} \div (11) = \{21 + 1\} \div 11 =$
$22 \div 11 = 2$

29. $(-2m^5 - 7m^3 - 5) + (-5m^5 - 2m^3) =$
$-2m^5 - 7m^3 - 5 + 5m^5 - 2m^3 =$
$3m^5 - 9m^3 - 5$

30. $(8x^3)(4xy^2) = 32x^4y^2$

31. $(-3x^3y)^2 = 9x^6y^2$

32. $(x + 2)(x + 1) =$
$x^2 + x + 2x + 2 = x^2 + 3x + 2$

33. $(4a - 2)(a + 1) =$
$4a^2 + 4a - 2a - 2 = 4a^2 + 2a - 2$

34. $2x^3 + 3x = x(2x^2 + 3)$

$2 \cdot x \cdot x \cdot x \quad 3 \cdot x$

35. $(2x)^0 = 1$

36. $\dfrac{x^6}{x^5} = x^1 = x$

37. $\dfrac{k^{10}m^{14}}{k^8} = k^2m^{14}$

38. $\dfrac{3x^2}{18} \cdot \dfrac{6}{x^3} = \dfrac{1}{x}$

39. $\dfrac{3}{4} \div \dfrac{9}{x} = \dfrac{3}{4} \cdot \dfrac{x}{9} = \dfrac{x}{12}$

40. $\dfrac{7}{12n^2} \div \dfrac{35}{n^3} = \dfrac{7}{12n^2} \cdot \dfrac{n^3}{35} = \dfrac{n}{60}$

41. $\dfrac{181y}{11} + \dfrac{27y}{11} = \dfrac{208y}{11}$

42. $\dfrac{3a}{5} + \dfrac{a}{7} = \dfrac{21a}{35} + \dfrac{5a}{35} = \dfrac{26a}{35}$

43. $\dfrac{4}{x^2} + \dfrac{6}{x^3} = \dfrac{4x}{x^3} + \dfrac{6}{x^3} = \dfrac{4x + 6}{x^3}$

44. $\dfrac{1}{a} + \dfrac{1}{b} = \dfrac{b}{ab} + \dfrac{a}{ab} = \dfrac{b + a}{ab}$

45. $23 + a = \dfrac{36}{6}$
$23 + a = 6$
$-23 \qquad = -23$
$\qquad a = -17$

46. $x + 8 - 3 = 5 + x$
$\quad x + 5 = 5 + x$
$\qquad\quad -5 = -5$
$\qquad\quad x = x$

any real number

76

47.
$$-\frac{1}{5}y = 3$$
$$-\frac{5}{1} \cdot -\frac{1}{5}y = 3 \cdot -\frac{5}{1}$$
$$y = -15$$

48.
$$7x = 56$$
$$\frac{1}{7} \cdot 7x = 56 \cdot \frac{1}{7}$$
$$x = 8$$

17.   1754.0000
             3.0001
       +   23.9000
          1780.9001

18.   Add the negative

             2.300
       +   5.006
       −   7.306 subtract small absolute
           7.306 value from the larger
           0.000

       0

19.   4.5300
          8.6651
       + 0.0090
       − 13.2041

20.   5.44763
          0.90000
          6.99900
        267.00000
        280.34663

21.   8.9000
          9.5600
         45.0000
       + 13.8476
         77.3076

22.   76.800
           7.680
            .768
       + 768.000
         853.248

23.   23.400
          46.390
       +   7.869
          77.659

24.   154.60
       +   78.93
           233.53

25.   43.90
             8.64
            52.54

          − 52.54

26.   − 0.64
          − 0.21
          − 0.850
            .364
          − 0.486

27.   − 532.000
          −   2.004
          − 534.004
               4.300
          − 529.704

28.   − 5.03
          − 0.36
          − 5.39

            82.30
          − 5.39
            76.91

29.   85.46
          − 71.21
           14.25

30.   984.87
           513.43
           471.44

31.   76.36
          − 48.45
           27.91

32. 823.96
     598.48
     225.48

33. 7.53 − 9.24 = 7.53 + (−9.24)

     9.24
     − 7.53
     − 1.71

34. (−3.65) − 2.14 = (−3.65) + (−2.14)

     3.65
     2.14
     − 5.79

35. 9.800
     − 8.957
     0.843

36. 7.363 − 9.8 = 7.363 + (−9.8)

     9.800
     7.363
     − 2.437

37. 24.45 − (−4.768) = 24.45 + 4.768

     24.450
     4.768
     29.218

38. − 10 − 8.76 = − 10 + (−8.76)

     10.00
     8.76
     − 18.76

39. −78.4 − 2.34 = −78.4 + (−2.34)

     78.40
     2.34
     − 80.74

40. 40.04
     4.40
     35.64

41. 70.005
     2.350
     67.655

42. 100.00
     3.08
     96.92

43. 75 − (−68.89) = 75 + 68.89

     75.00
     68.89
     143.89

44. 3.473 − (−24.47) = 3.473 + 24.47

     3.473
     24.470
     27.943

45. −7.54 − (−7.54) = −−7.54 + 7.54 = 0

46. 657.874
     − 472.097
     185.777

47. 905.008
     − 799.999
     105.009

48. 460.700
     − 4.607
     456.093

49. 46.36 − 782.3 = 46.36 + (−782.3)

     782.30
     46.36
     − 735.94

50. $-76.3 - (-89) = -76.3 + 89$

$$\begin{array}{r} 89.0 \\ -\ 76.3 \\ \hline 12.7 \end{array}$$

51. $-3.86 - 93.4 = -3.86 + (-93.4)$

$$\begin{array}{r} 3.86 \\ 93.40 \\ \hline -\ 97.26 \end{array}$$

52. $$\begin{array}{r} 782.34 \\ -\ 764.80 \\ \hline 17.54 \end{array}$$

53. $$\begin{array}{r} 100.000 \\ -\ \ \ 0.863 \\ \hline 99.137 \end{array}$$

54. $-100 - 0.863 = -100 + (-0.863)$

$$\begin{array}{r} 100.000 \\ -\ \ \ 0.863 \\ \hline -\ 100.863 \end{array}$$

55. $9.7 - 99.34 = 9.7 + (-99.34)$

$$\begin{array}{r} 99.34 \\ 9.70 \\ \hline -\ 89.64 \end{array}$$

56. $-6.4 - 63.54 = -6.4 + (-63.54)$

$$\begin{array}{r} 6.40 \\ 63.54 \\ \hline -\ 69.94 \end{array}$$

57. $4.2 + (6.8 - 2.4) = 4.2 + 4.4 = 8.6$

58. $(9.3 - 8.5) + 4.7 = 0.8 + 4.7 = 5.5$

59. $6.4 - (3.2 + 5.6) = 6.4 - 8.8 = -2.4$

60. $(6.77 + 3.53) - 11 = 10.3 - 11 = -0.7$

61. $3.45 - (5.43 - 4.3) = 3.45 - 1.13 = 2.32$

62. $(9.75 - 3.5) - 1.2 = 6.25 - 1.2 = 5.05$

63. $(36.4 - 76.4) + 2.6 = -40 + 2.6 = -37.4$

64. $9.6 + (-99.7) = 9.6 + (-13.4) = -3.8$

65. $(4.8 - 9.46) + (3.8 - 1.29) =$
$(-4.66) + (2.51) = -2.15$

66. $(32.1 + 4.82) - (4.61 + 28.3) =$
$(36.92) - (32.91) = 4.01$

67. $100 - [76.8 - (-8.63)] =$
$100 - [76.8 + 8.63] =$
$100 - (85.43) = 14.57$

68. $(546.3 - 2.96) - 836.54 =$
$546.34 - 836.54 = -293.2$

69. $(1.864 - 32.7) + 3.86 =$
$(-30.386) + 3.86 = -26.976$

70. $0.346 - (0.46 - 0.973) =$
$0.346 - (-0.513) =$
$0.346 + 0.513 = 0.859$

71. $9.46 + 5.39 = 14.85$

72. $8.9 - 4.56 = 4.34$

73. $10.4 - 7.332 = 3.068$

74. $(5.2 + 4.876) - 3.98 =$
$10.076 - 3.98 = 6.096$

75. $9.86 + 32.7 = 42.56$

76. $932.4 + 72.63 = 1005.27$

77. $8.93 - 47.3 = 8.93 + (-47.3) = -38.37$

80

78. $3.86 - 0.049 = 3.811$

79.
```
   2.80
   7.49
   0.36
+ 296.00
 306.65
```

80.
```
  12.30
-  9.63
   2.67
```

81.
```
  6.004
- 2.380
  3.624
```

82. $2.83 - 8.46 = 2.83 + (-8.46)$

```
   8.46
   2.83
-  5.63
```

83.
```
   35.59
   47.98
   54.45
    7.99
    4.68
+  10.50
```
$161.67 spent on text books.

84.
```
  23.59
  19.97
  20.07
  45.50
```
$109.13 spent on holiday shopping.

85.
```
  5.48819
- 4.38970
  1.09849 minutes faster.
```

86.
```
   175.75
    54.67
+   32.65
```
$263.07 bills paid.

```
  306.94
- 263.07
```
$ 43.87 left in checking account.

87.
```
   14.50
   17.86
+  22.50
   54.86 square yards needed.
-  50.00
    4.86
```
1 bolt and 4.86 yd^2 extra needed.

88.
```
   3.52
   2.98
+  4.90
  11.40    11.4 centimeters
```

89.
```
   8.64
   8.64
   3.96
   3.96
  25.20    25.2 inches
```

90. $2\frac{1}{2} + 2\frac{1}{2} + 2\frac{1}{2} + 2\frac{1}{2} =$

$2 + 2 + 2 + 2 + \frac{1}{2} + \frac{1}{2} + \frac{1}{2} + \frac{1}{2} =$

$8 + \frac{4}{2} = 8 + 2 = 10$ feet.

7.3 Multiplication of Decimals

1. 0.9(0.7) $9 \times 7 = 63$
   = 0.63

2. (0.6)(0.5) $6 \times 5 = 30$
   = 0.30 = 0.3

3. 1.1(0.9) $11 \times 9 = 99$
   = 0.99

4. (0.4)0.8 $4 \times 8 = 32$
   = 0.32

5. (0.5)(-0.4) $4 \times 5 = 20$
   = -0.20 = -0.2

6. (0.3)(0.8) $3 \times 8 = 24$
   = 0.24

7. (0.3)(0.8) $3 \times 8 = 24$
   = 0.024

8. (0.003)(0.8) = 0.0024
9. (0.3)(0.08) = 0.024
10. (0.3)(0.008) = 0.0024
11. (0.03)(0.08) = 0.0024
12. (0.003)(0.08) = 0.00024
13. (0.03)(0.008) = 0.00024

14. (-0.6)(0.9) $6 \times 9 = 54$
   = -0.54

15. (-0.6)(0.7) $6 \times 7 = 42$
   = 0.42

16. 0.0008(0.008) = 0.0000064
17. (-0.0009)(0.00001) = -0.000000009
18. (-0.007)(-0.003) = 0.000021
19. 231:352 (1.5) = 347.028
20. 78.59(36.1) = 2837.099
21. 92.7(-0.007) = -0.6489
22. 67.8(0.7) = 47.46
23. 67.8(0.07) = 4.746

24. 67.8(0.007) = 0.4746
25. 67.8(0.0007) = 0.04746

26. -43.7(100) = -4370
    Move the decimal point 2 places to the right.

27. 75.43(-100) = -7543
    Move the decimal point 2 places to the right.

28. 5.003(10) = 50.03
    Move the decimal point 1 place to the right.

29. (-98.999)(-1000) = 98,999
    Move the decimal point 3 places to the right.

30. 6.7(1000) = 6700
    Move the decimal point 3 places to the right.

31. 5.42(-10) = -54.2
    Move the decimal point 1 place to the right.

32. (0.00005)10 = 0.0005
    Move the decimal point 1 place to the right.

33. (0.00005)100 = 0.005
    Move the decimal point 2 places to the right.

34. (0.00005)1000 = 0.05
    Move the decimal point 3 places to the right.

35. (0.00005)10,000 = 0.5
    Move the decimal point 4 places to the right.

36. 7.2 + (0.2)(0.4) = 7.2 + 0.08 = 7.28

37. $5.34(0.01) + 21.87 =$
    $0.0534 + 21.87 = 21.9234$

38. $[2.1 + 3.7][3.3 - 5.7] =$
    $[5.8][-2.4] = -13.92$

39. $0.9(2.5 + 5.2) = 0.9(7.7) = 6.93$

40. $-0.3(-7.6 - 2.4) = -0.3(-10) = 3$

41. $[2.5 - 9.6][-7.6 + 3.9] =$
    $[-7.1][-3.7] = 26.27$

42. $9.5008 + (0.3)(-0.7) =$
    $9.5008 + (-0.21) = 9.2908$

43. $(7.5 + 3.1) + (8.6 - 4.2) =$
    $(10.6) + (4.4) = 15$

44. $(7.003 - 9.1)(3.001) =$
    $(-2.097)(3.001) = -6.293097$

45. $(9.2 - 0.003) - (8.6 - 9.2) =$
    $(9.197) - (-0.6) =$
    $9.197 + 0.6 = 9.797$

46. $7.41(8.29 - 12.006) =$
    $7.41(-3.716) = -27.53556$

47. $(3.86 + 2.93)(2.1 - 6.89) =$
    $(6.79)(-4.79) = -32.5241$

48. $7.2(8.23) - 2.6(3.92) =$
    $59.256 - 10.192 = 49.064$

49. $(6.323)(0.6) = 3.738$

50. $(0.004)(0.0029) = 0.0000116$

51. $7(0.0086) = 0.0602$

52. $2(8.375) = 16.75$

53. $3(7.89 + 3.2) = 3(11.09) = 33.27$

54. $(3.89)(6.2) = 24.118$

55. $(3)(0.002)(7.8) = 0.0468$

56. $(5.32 + 3.64)4 = (8.96)4 = 35.84$

57. $2.5 + (5.8)(2.1) = 2.5 + 12.18 = 14.68$

58. 15 boxes at 0.35 each
    $15(0.35) = \$5.25$

    ```
      5.25
    - 4.99
       .26
    ```

    The box is cheaper by 26 cents.

59. 82 windows using 7.83 meters
    $82(7.83) = 642.06$ meters

    642.06 meters at $4.99
    $(642.06)(4.99) = 3203.88$

    $3,203.88 cost of fabric

60. 30 cents first minute + 26 cents for each
    of 12 minutes
    $0.30 + (0.26)(12) = 0.30 + (3.12) = 3.42$
    $3.42 for a 13-minute call.

61. 21 commercials at 1.42 minutes each
    $21(1.42) = 29.82$ minutes
    2 hours = 120 minutes
    $120 - 29.82 = 90.18$
    The movie ran 90.18 minutes.

62. 8 bars at 42 cents each
    $8(0.42) = 3.36$
    $3.36 - 2.99 = 0.37$
    The box is a better buy saving 37 cents.

63.   17 windows each needing 14.85 yards
       17(14.85) = 252.45 yards
       252.45 yards at $8.95 per yard
       (252.45)(8.95) = 2,259.4275
       Fabric will cost $2.259.43.

64. $2.50 for first 3 minutes + 85 cents for
       each of 16 minutes.
       2.5 + (0.85)(16) = 2.50 + 13.60 = 16.10
       $16.10 for a 19-minute call.

65. $2.95 for the first 3 minutes + 87 cents
       for each of 20 minutes.
       2.95 + (0.87)(20) =
       2.95 + 17.40 = 20.35
       $20.35 for a 23-minute call.

66.   6 pairs at $2.03 per pair
       6(2.03) = $12.18
       12.18 − 11.99 = 0.19
       The pack is cheaper by 19 cents.

67. $2 to pick up + 75 cents for each of
       4 miles.
       2 + (0.75)(4) = 2 + 3 = 5
       $5 taxi fare.

68. $2 to pick up and 70 cents for each
       ½ mile
       9 miles is 18 -- ½ miles
       2 + (0.70)(18) = 2 + 12.6 = 14.6
       $14.60 taxi fare.

69.   7.82(4) = 31.28 inches

7.4 Division of Decimals

1. $0.534 \div 0.3 = 1.78$

$$
\begin{array}{r}
1.78 \\
0.3\,\overline{)\,0.534} \\
\underline{3} \\
23 \\
\underline{21} \\
24 \\
24
\end{array}
$$

2. $19.888 \div 0.8 = 24.86$

3. $408.03 \div 7 = 58.29$

4. $938.05 \div 5 = 187.61$

5. $15.0004 \div 0.05 = 300.008$

$$
\begin{array}{r}
300.008 \\
0.05\,\overline{)\,15.00040} \\
\underline{15} \\
000040 \\
40
\end{array}
$$

6. $1.8123 \div 0.06 = 30.205$

7. $8 \div 0.2 = 40$

$$
\begin{array}{r}
40. \\
0.2\,\overline{)\,8.0} \\
\underline{8}
\end{array}
$$

8. $0.8 \div 2 = 0.4$

$$
\begin{array}{r}
0.4 \\
2\,\overline{)\,0.8}
\end{array}
$$

9. $0.8 \div 0.2 = 4$

$$
\begin{array}{r}
4. \\
0.2\,\overline{)\,0.8}
\end{array}
$$

84

10. $0.08 \div 0.2 = 0.4$

$$0.2\overline{)0.08}\;\;^{.4}$$

11. $0.008 \div 0.2 = 0.04$

$$0.2\overline{)0.008}\;\;^{.04}$$

12. $0.0008 \div 0.2 = 0.004$

$$0.2\overline{)0.0008}\;\;^{.004}$$

13. $0.00008 \div 0.2 = 0.0004$

$$0.2\overline{)0.00008}\;\;^{.0004}$$

14. $0.8 \div 0.02 = 40$

$$0.02\overline{)0.80}\;\;^{40.}$$

15. $0.8 \div 0.002 = 400$

$$0.002\overline{)0.800}\;\;^{400.}$$

16. $0.8 \div 0.0002 = 4,000$

$$0.0002\overline{)0.8000}\;\;^{4000.}$$

17. $0.8 \div 0.00002 = 40,000$

$$0.00002\overline{)0.80000}\;\;^{40000.}$$

18. $-55.368 \div 2.4 = -23.07$

19. $1595.139 \div (-3.9) = -409.01$

20. $-71.064 \div 7.2 = -9.87$

21. $206.55 \div 4.5 = 45.9$

22. $-0.7921 \div (-0.89) = 0.89$

23. $145.8 \div 0.36 = 405$

24. $101.455 \div 1.03 = 98.5$

25. $2372.49 \div (-2.61) = -909$

26. $-13208.08 \div 21.4 = --617.2$

27. $46.83 \div 10 = 4.683$
Move the decimal point 1 place to the left.

28. $0.864 \div 10 = 0.0864$
Move the decimal point 1 place to the left.

29. $8536 \div 100 = 85.35$
Move the decimal point 2 places to the left.

30. $793.8 \div 100 = 7.938$
Move the decimal point 2 places to the left.

31. $53864.6 \div 1000 = 53.8646$
Move the decimal point 3 places to the left.

32. $2.587 \div 0.7 = 3.70$

33. $33.39 \div 0.4 = 83.48$

34. $-128.9 \div .6 = -214.83$

35. $0.065 \div -0.8 = -0.08$

36. $9.8 \div 0.13 = 75.38$

37. $54.7 \div (-2.3) = -23.78$

38. $97.6 \div 17 = 5.74$

39. $72.7 \div 1.9 = 38.26$

40. $0.633 \div 5.7 = 0.11$

41. $749 \div 29 = 25.83$

42. $0.0035 \div 2.56 = 0.00$

43. $4.23 \div 3.8 = 1.11$

44. $672 \div 0.82 = 819.51$

45. $7.638 \div 7 = 1.09$

46. $0.049 \div 7 = 0.01$

47. $2.38 \div 4 = 0.595$

48. $22.7737 \div 2.9 = 7.853$

49. $54.33953 \div 7.83 = 6.94$

50. Brand A costs $3.22 for 14 oz.
$3.22 \div 14 = 0.23$ or 23 cents per oz.
Brand B costs $4.18 for 19 oz.
$4.18 \div 19 = 0.22$ or 22 cents per oz.

 Brand B is a better buy saving 1 cent per oz.

51. 32 bars cost $10.88
$10.88 \div 32 = 0.34$
Each bar costs 34 cents.

52. 1 dozen or 12 pencils cost $2.52
$1.52 \div 12 = 0.21$
Each pencil costs 21 cents.

53. Books cost $ 54.99
                 $ 67.42
                 $ 59.00
                 $181.41

 $181.41 \div 3 = 60.47$
Each student will pay $60.47.

54. Call costs $11.12.
First minute costs 85 cents.
$11.12 - 10.27$
$10.27 \div 79$ cents cost per minute =
13 minutes, $13 + 1 = 14$
A 14-minute call

55. 15 tickets cost $126.45
$126.45 \div 15 = 8.43$
Each ticket costs $8.43.

56. Cleano costs $2.36 for 29.5 oz.
$2.36 \div 31.5 = 0.08$
Cleano costs 8 cents per oz.

 Betto costs $1.89 for 31.5 oz.
$1.89 \div 31.5 = 0.06$
Betto costs 6 cents per oz.

 Betto is a better buy costing 2 cents an oz. Less than Cleano.

57. $9.679 + 3.49 + 0.6325 = 13.80$

58. $5.432 + 4.32 + 3.2 + 2 + 0.23 = 15.18$

59. $0.396 \div (7 + 3.96) =$
$0.396 \div 10.96 = 0.036 \approx 0.04$

60. $796.45 - 32.682 = 763.768 \approx 763.77$

61. $5.07(-2.4 - 0.631) =$
$5.97(-3.031) = -18.095 \approx 18.10$

62. $(9.37 - 12.5)(-8) =$
$(-3.13)(-8) = 25.04$

86

63. $3.3 - 4(0.22) = 3.3 - 0.88 = 2.42$

64. $54.987 - 2(3.05 + 0.51) =$
$54.987 - 2(3.201) =$
$54.987 - 6.402 = 48.585 \approx 48.59$

65. $54.987 - 2(-3.05 + 0.51) =$
$54.987 - 2(-2.899) =$
$54.987 + 5.798 = -49.189 \approx -49.19$

66. $36.01 \div (4.101 - 1.01) =$
$36.01 \div 3.091 = 11.65$

67. $(2.3 + 1.2) \div 4 = 3.5 \div 4 = 0.875 \approx 0.88$

68. $(7.6 \div 0.2)3.2 = (38)3.2 = 121.6$

69. $(3.04 + 2.66) \div (0.34 - 2.34) =$
$5.7 \div (-2) = -2.85$

70. $9.8(12.63 - 128.9) =$
$9.8(-116.27) = -1139.446 \approx 1139.45$

71. $(7.01)(9.63) \div 2.8 =$
$67.5063 \div 2.8 = 24.11$

72. $8.6 \div 4 = 2.15$
Each side has a length of 2.15 inches.

7.5 Conversions Between Fractions and Decimals

1. $\dfrac{1}{2} = 1 \div 2 = 0.5$

2. $\dfrac{1}{3} = 1 \div 3 = 0.\overline{3}$

3. $\dfrac{1}{4} = 1 \div 4 = 0.25$

4. $\dfrac{2}{5} = 2 \div 5 = 0.4$

5. $1\dfrac{1}{2} = 1.5$

The whole number is the same as a fraction or a decimal.

6. $\dfrac{7}{8} = 7 \div 8 = 0.875$

7. $\dfrac{2}{3} = 2 \div 3 = 0.\overline{6}$

8. $\dfrac{3}{4} = 3 \div 11 = 0.\overline{27}$

9. $3\dfrac{5}{6};$ the whole number is the same

$3; \dfrac{5}{6}$ is $5 \div 6 = 0.8\overline{3}$

$3\dfrac{5}{6} = 3.8\overline{3}$

10. $\dfrac{7}{5} = 7 \div 5 = 1.4$

11. $\dfrac{3}{7} = 3 \div 7 = 0.\overline{428571}$

12. $\dfrac{4}{9} = 4 \div 9 = 0.\overline{4}$

13. $2\dfrac{1}{12};$ the whole number is the same

$\dfrac{1}{12} = 1 \div 12 = 0.08\overline{3}$

$2\dfrac{1}{12} = 2.08\overline{3}$

14. $\dfrac{3}{8} = 3 \div 8 = 0.375$

15. $2\dfrac{1}{8} = 2 + 1 \div 8 = 2.125$

16. $4\dfrac{3}{4}$; the whole number is the same

   $\dfrac{3}{4} = 3 \div 4 = 0.75$

   $4\dfrac{3}{4} = 4.75$

17. $\dfrac{1}{20} = 1 \div 20 = 0.05$

18. $\dfrac{7}{100} = 0.07$

19. $\dfrac{3}{20} = 3 \div 20 = 0.15$

20. $\dfrac{61}{1000} = 0.061$

21. $\dfrac{4}{5} = 4 \div 5 = 0.8$

22. $\dfrac{9}{8} = 9 \div 8 = 1.125$

23. $\dfrac{3}{4} = 3 \div 4 = 0.75$

24. $\dfrac{3}{5} = 3 \div 5 = 0.6$

25. $\dfrac{1}{6} = 1 \div 6 = 0.1\overline{6}$

26. $\dfrac{5}{9} = 5 \div 9 = 0.\overline{5}$

27. $\dfrac{5}{6} = 5 \div 6 = 0.8\overline{3}$

28. $\dfrac{15}{16} = 15 \div 16 = 0.9375$

29. $\dfrac{3}{17} = 2 \div 17 =$

   $0.\overline{1176470588235294}$

30. $\dfrac{3}{40} = 3 \div 40 = 0.075$

31. $0.7 = \dfrac{7}{10}$

32. $0.35 = \dfrac{35}{100} = \dfrac{7}{20}$

33. $0.8 = \dfrac{8}{10} = \dfrac{4}{5}$

34. $0.5 = \dfrac{5}{10} = \dfrac{1}{2}$

35. $0.05 = \dfrac{5}{100} = \dfrac{1}{20}$

36. $0.005 = \dfrac{5}{100} = \dfrac{1}{200}$

37. $0.0005 = \dfrac{5}{10000} = \dfrac{1}{2000}$

38.
   $0.25 = \dfrac{25}{100} = \dfrac{1}{4}$

39. $2.75 = 2\dfrac{75}{100} = 2\dfrac{3}{4}$

40. $3.2 = 3\dfrac{2}{10} = 3\dfrac{1}{5}$

41. $0.015 = \dfrac{15}{1000} = \dfrac{3}{200}$

42. $0.625 = \dfrac{625}{1000} = \dfrac{25}{40} = \dfrac{5}{8}$

43. $0.07 = \dfrac{7}{100}$

44. $0.065 = \dfrac{65}{1000} = \dfrac{13}{200}$

45. $0.012 = \dfrac{12}{1000} = \dfrac{3}{250}$

46. $0.17 = \dfrac{17}{100}$

47. $0.34 = \dfrac{34}{100} = \dfrac{17}{50}$

48. $0.875 = \dfrac{875}{1000} = \dfrac{34}{40} = \dfrac{7}{8}$

49. $3.125 = 3\dfrac{125}{1000} = 3\dfrac{5}{40} = 3\dfrac{1}{8}$

50. $0.0002 = \dfrac{2}{10000} = \dfrac{1}{5000}$

51. $3.12 = 3\dfrac{12}{100} = 3\dfrac{3}{25}$

52. $1.45 = 1\dfrac{45}{100} = 1\dfrac{9}{20}$

53. $236 \div 236$

54. $4.5 = 4\dfrac{5}{10} = 4\dfrac{1}{2}$

55. $7.004 = 7\dfrac{4}{1000} = 7\dfrac{1}{250}$

56. $0.2 = \dfrac{2}{10} = \dfrac{1}{5}$

57. $0.02 = \dfrac{2}{100} = \dfrac{1}{50}$

58. $0.002 = \dfrac{2}{1000} = \dfrac{1}{500}$

59. $0.0002 = \dfrac{2}{10000} = \dfrac{1}{5000}$

60. $0.00002 = \dfrac{2}{100000} = \dfrac{1}{50000}$

61. $\dfrac{3}{4} + 0.5 = 0.75 + 0.5 = 1.25$

62. $\dfrac{3}{4} - 0.5 = 0.75 - 0.5 = 0.25$

63. $\dfrac{1}{3} + 0.25 = \dfrac{1}{3} + \dfrac{1}{4} =$

    $\dfrac{4}{12} + \dfrac{3}{12} = \dfrac{7}{12} = 0.58\overline{3}$

64. $\dfrac{1}{3} - 0.25 = \dfrac{1}{3} - \dfrac{1}{4} =$

    $\dfrac{4}{12} - \dfrac{3}{12} = \dfrac{1}{12} = 0.08\overline{3}$

65. $0.8 - \dfrac{4}{5} = 0.8 - 0.8 = 0$

66. $3\dfrac{1}{2} \cdot (0.25) = 3.5(0.25) = 0.875$

67. $\dfrac{7}{8}(2.4) = 0.875(2.4) = 2.1$

68. $(0.04)(\frac{3}{4} + 0.25) =$

$0.04(0.75 + 0.25) = 0.04(1) = 0.04$

69. $(\frac{3}{8} - 0.125)\frac{2}{3} =$

$(\frac{3}{8} - \frac{1}{8})\frac{2}{3} = \frac{2}{8} \cdot \frac{2}{3} =$

$\frac{1}{6} = 0.1\overline{6}$

70. $1.7(2 - \frac{3}{17}) = \frac{17}{10}(\frac{34}{17} - \frac{3}{17}) =$

$\frac{17}{10} \cdot \frac{31}{17} = \frac{31}{10} = 3.1$

71. $\frac{1}{3} + 0.6(\frac{1}{3}) = \frac{1}{3} + \frac{6}{10} \cdot \frac{1}{3}$

$\frac{1}{3} + \frac{1}{5} = \frac{5}{15} + \frac{3}{15} =$

$\frac{8}{15} = 0.5\overline{3}$

72. $\frac{3}{5} + 0.7(\frac{4}{7}) = \frac{3}{5} + \frac{7}{10} \cdot \frac{4}{7}$

$\frac{3}{5} + \frac{2}{5} = \frac{5}{5} = 1$

73. $\frac{1}{3}(1.2) + 0.5(\frac{2}{3}) =$

$(\frac{1}{3} \cdot \frac{12}{10}) + (\frac{1}{2} \cdot \frac{2}{3}) = \frac{2}{5} + \frac{1}{3}$

$\frac{6}{15} + \frac{5}{15} = \frac{11}{15} = 0.7\overline{3}$

74. $0.75(1\frac{1}{3}) + \frac{5}{8}(0.25) =$

$(\frac{3}{4} \cdot \frac{4}{3}) + (\frac{5}{8} \cdot \frac{1}{4}) = 1 + \frac{5}{32} =$

$1\frac{5}{32}$ or 1.15625

75. $\frac{2}{5}(0.86 + 2.83) = 0.4(3.69) = 1.476$

76. $(\frac{3}{4} - 0.125)(0.5 - \frac{3}{8}) =$

$(0.75 - 0.125)(0.5 - 0.375) =$

$(0.625)(0.125) = 0.078125$

77. $-1 - 4.25 + \frac{34}{8} =$

$-1 + (-4.25) + 4.25 = -1$

78. $-2 - 3.75 + \frac{15}{4} =$

$-2 + (-3.75) + 3.75 = -2$

79. $0.9 - \frac{9}{25} + 0.18 =$

$0.9 - 0.36 + 0.18 = 0.72$

90

80. $5 - 0.5 + \dfrac{5}{8} =$

$5 + (-0.5) + 0.625 = 5.125$

81. $0.875 + \dfrac{5}{6} + \dfrac{1}{12} =$

$\dfrac{7}{8} + \dfrac{5}{6} + \dfrac{1}{12} = \dfrac{21}{24} + \dfrac{20}{24} + \dfrac{2}{24}$

$\dfrac{43}{24} = 1\dfrac{19}{24}$ or $1.791\overline{6}$

82. $0.75 - \dfrac{1}{2} + 0.25 + 5\dfrac{1}{2} =$

$0.75 + (-0.5) + 0.25 + 5.5 = 6$

83. $-\dfrac{1}{2} + 1.25 - \dfrac{7}{8} - 0.5 =$

$-0.5 + 1.25 + (-0.875) + (-0.5) =$

-0.625

84. $0.75 - \dfrac{1}{5} + 0.1 =$

$0.75 + (-0.2) + 0.1 = 0.65$

85. $\dfrac{1}{3} + 5 - 0.2 = \dfrac{1}{3} + 5 + (-\dfrac{1}{5}) =$

$5 + \dfrac{5}{15} + (-\dfrac{3}{15}) = 5 + \dfrac{2}{15} = 5\dfrac{2}{15}$

86. $0.9 - \dfrac{1}{2} + 0.25 =$

$0.9 + (-0.5) + 0.25 = 0.65$

87. $2.5 + 7 + 0.4 - 20 =$

$2.5 + 7 + 0.4 + (-20) = -10.1$

88. $0.875 + \dfrac{3}{8} + 0.875 + \dfrac{3}{8} =$

$0.875 + 0.375 + 0.875 + 0.375 =$

2.5 inches or $2\dfrac{1}{2}$ inches

89. $1.125 + \dfrac{1}{8} + \dfrac{7}{8} =$

$1.125 + 0.125 + 0.875 =$

2.125 inches or $2\dfrac{1}{8}$ inches

90. $(2.75)4 = 11$ inches

$(2\dfrac{3}{4})4 = \dfrac{11}{4} \cdot \dfrac{4}{1} = 11$ inches

REVIEW

1. Decimals, see Section 7.1
2. Irrational Numbers, see Section 7.1
3. Scientific Notation, see Section 7.1
4. Standard Notation, see Section 7.1
5. 4.653, hundredths
6. 76.5, tenths
7. 0.0005, ten thousandths
8. 5.36, units
9. 946.753, hundredths
10. 23.24509, thousandths
11. 0.0000005, ten millionths
12. 53, tens

91

13.	$0.06 < 0.6$	35.	$7,860,000 = 7.86 \times 10^6$
14.	$0.3 < 0.\overline{3}$	36.	$3,000 = 3 \times 10^3$
15.	$-0.6 > -0.65$	37.	$0.00000098 = 9.8 \times 10^{-7}$
16.	$1.7 > 1.17$	38.	$0.009684 = 9.684 \times 10^{-3}$
17.	$0.0004 < 0.004$		
18.	$-0.72 > -0.73$	39.	36.279 to the nearst tenth, 36.3
19.	$9.01 < 9.1$	40.	793.83 to the nearest ten, 790
20.	$0.0006 > 0.00006$	41.	0.83709 to the nearest thousandth, 0.837

21. Twenty-seven and twenty-seven thousandths. 27.027

22. Four hundred and seven hundredths. 400.07

23. Five thousand, and five thousandths. 5,000.005

24. Three hundred two and two hundredths. 302.02

25. Seven hundred sixty-three thousandths. 0.763

26. 75.75   Seventy-five and seventy-five hundredths.

27. 930.003   Nine hundred thirty and three thousandths.

28. 4.5002   Four and five thousand two ten-thousandths.

29. 0.00678   Six hundred seventy-eight hundred thousandths.

30. 12.4004   Twelve and four thousand four ten-thousandths.

31. $6.73 \times 10^6 = 7,730,000$
32. $2.4 \times 10^{-5} = 0.000024$
33. $6.08 \times 10^{-3} = 0.00608$
34. $7 \times 10^9 = 7,000,000,000$

42. 99.999 to the nearest hundredth, 100.00
43. 99.999 to the nearest hundred, 100
44. 700.007 to the nearest tenth, 700.0

45.
```
    2.83
   17.40
 +362.00
  382.23
```

46.
```
   0.3
   3.00
 + 3.03
   6.33
```

47.
```
   3.6480
 + 0.9376
   4.5856
 - 2.0800
   2.5056
```

48.
```
     0.003
 + 300.000
   300.003
     3.030
   296.973
```

49.
```
   -45.2
    -3.5
 + -628.0
  -676.7
```

50.
```
   342.090
     0.261
   342.351
 -  71.309
   271.042
```

51.
```
    23.33        -456.00
    86.90      - +110.23
   110.23        -345.77
```

52.
```
     0.009
     0.981
   972.000
   972.990
```

53.
```
     9.630
   - 6.897
     2.733
```

54.
```
    26.30
  -  9.37
    16.93
```

55.
```
     -7.84
  + -83.70
    -91.54
```

56. $-15.2 - (-2.38) = 15.2 + 2.38$
```
    -15.20
  - +2.38
    -12.82
```

57.
```
   236.720
  - 23.672
   213.048
```

58. $7.86 - 19.4 = 7.86 + (-19.4)$
```
   -19.40
  - +7.86
   -11.54
```

59. $-3.28 - (-9.6) = -3.28 + 9.6$
```
   +9.60
   -3.28
    6.32
```

60. $-6.73 - 6.73 = -6.73 + (-6.73)$
```
     -6.73
   + -6.73
    -13.46
```

61. (0.3)1.2   12 x 3 = 36
     = 0.36

62. (56.1)(0.003)   561 x 3 = 1683
     = 0.1683

63. (0.0005)(0.0004)   5 x 4 = 20
     = 0.00000020 = 0.0000002

64. (-2.7)(4.03)   27 x 403 = 10881
     = -10.881

65. (-9.32)(-0.11)   932 x 11 = 10252
     = 1.0252

66. (39.87)(1000) = 39870
     Move the decimal point 3 places to
     the right.

67. 0.0098(100) = 0.98
     Move the decimal point 2 places to
     the right.

68. 79.00987(1000) = 79009.87
     Move the decimal point 3 places to
     the right.

69. $78.6 \div 3 = 26.2$

$$
\begin{array}{r}
26.2 \\
3\overline{\smash{)}78.6} \\
\underline{6} \\
18 \\
\underline{18} \\
06 \\
\underline{6}
\end{array}
$$

70. $3.79 \div 0.05 = 75.8$

$$
\begin{array}{r}
75.8 \\
0.05\overline{\smash{)}3.790} \\
\underline{35} \\
29 \\
\underline{25} \\
40 \\
\underline{40}
\end{array}
$$

71. $1.88 \div 8 = 0.235 = 0.24$

$$
\begin{array}{r}
.235 \\
8\overline{\smash{)}1.880} \\
\underline{16} \\
28 \\
\underline{24} \\
40 \\
40
\end{array}
$$

72. $94 \div 5 = 18.8$

$$
\begin{array}{r}
18.8 \\
5\overline{\smash{)}94.0} \\
\underline{5} \\
44 \\
\underline{40} \\
40 \\
40
\end{array}
$$

73. $16.64 \div 3.2 = 5.2$

$$
\begin{array}{r}
5.2 \\
3.2\overline{\smash{)}16.64} \\
\underline{160} \\
64 \\
64
\end{array}
$$

74. $7.26 \div 1.8 = 4.0\overline{3}$

$$
\begin{array}{r}
4.0333 \\
1.8\overline{\smash{)}7.2600} \\
\underline{72} \\
060 \\
\underline{54} \\
60 \\
\underline{54}
\end{array}
$$

75. $74.83 \div 1000 = 0.07483 - 0.97$
Move the decimal point 3 places to the left.

76. $238.976 \div 100 = 2.38976 = 2.39$
Move the decimal point 2 places to the left.

77. $0.009 \div 1000 = 0.000009$
Move the decimal point 3 places to the left.

78. $\dfrac{4}{5} = 4 \div 5 = 0.8$

79. $\dfrac{2}{3} = 2 \div 3 = 0.\overline{6}$

80. $\dfrac{2}{1} = 2 \div 11 = 0.\overline{18}$

81. $\dfrac{7}{8} = 7 \div 8 = 0.875$

82. $\dfrac{5}{2} = 5 \div 2 = 2\dfrac{1}{2} = 2.5$

94

83. $\dfrac{43}{100} = 0.43$

84. $\dfrac{7}{100} = 0.07$

85. $\dfrac{3}{8}$ $3 \div 8 = 0.375$

86. $0.25 = \dfrac{25}{100} = \dfrac{1}{4}$

87. $2.125 = 2\dfrac{125}{1000} = 2\dfrac{1}{8}$

88. $0.005 = \dfrac{5}{1000} = \dfrac{1}{200}$

89. $0.61 = \dfrac{61}{100}$

90. $0.375 = \dfrac{375}{1000} = \dfrac{3}{8}$

91. $1.75 = 1\dfrac{75}{100} = 1\dfrac{3}{4}$

92. $0.875 = \dfrac{875}{1000} = \dfrac{7}{8}$

93. $3.5 = 3\dfrac{5}{10} = 3\dfrac{1}{2}$

94. $\dfrac{4}{3} - \dfrac{1}{6} + 0.75 = \dfrac{4}{3} - \dfrac{1}{6} + \dfrac{3}{4} =$

$\dfrac{16}{12} - \dfrac{2}{12} + \dfrac{9}{12} = \dfrac{23}{12} = 1\dfrac{11}{12}$

95. $\dfrac{7}{8} + 1\dfrac{1}{2}(0.125 + \dfrac{3}{4}) =$

$0.875 + 1.5(0.125 + 0.75) =$
$0.875 + 1.5(0.875) =$
$0.875 + 1.3125 = 2.1875$

96. $0.5 - 2\dfrac{2}{3}(0.25 + \dfrac{3}{4}) =$

$\dfrac{1}{2} - 2\dfrac{22}{3}(\dfrac{1}{4} + \dfrac{3}{4}) = \dfrac{1}{2} - 2\dfrac{2}{3}(1) =$

$\dfrac{1}{2} - 2\dfrac{2}{3} = \dfrac{1}{2} + (-2\dfrac{2}{3}) =$

$\dfrac{1}{2} + (-\dfrac{8}{3}) = \dfrac{3}{6} + (-\dfrac{16}{6}) =$

$-\dfrac{13}{6} = -2\dfrac{1}{6}$

97. $(\dfrac{1}{6} + 0.5)(4 + 0.375) =$

$(\dfrac{1}{6} + \dfrac{1}{2})(4 + \dfrac{3}{8}) =$

$(\dfrac{2}{12} + \dfrac{6}{12})(\dfrac{32}{8} + \dfrac{3}{8}) = \dfrac{8}{12} \cdot \dfrac{35}{8}$

$\dfrac{35}{12} = 2\dfrac{11}{12}$

98. $1 - 2.75 + \dfrac{11}{4} =$

$1 - 2.75 + 2.75 = 1$

99. $0.6 + \dfrac{7}{10} + 6.1 =$

$0.6 + 0.7 + 6.1 = 7.4$

100. $0.3 - 8 - 9 + 0.2 =$

$0.3 + (-8) + (-9) + 0.2 = -16.5$

101. $18 - 0.4 + 2.6 - 3 =$

$18 + (-0.4) + 2.6 + (-3) = 17.2$

PRACTICE TEST

1. tenths place is 0
   hundredths place is 4

2. $0.6 = 0.600$
   $0.\overline{6} = 0.666....$
   $\quad 0.\overline{6}$ is larger

3. $0.005 = \dfrac{5}{1000} = \dfrac{1}{200}$

4. $-0.05 = -0.050$
   $-0.005 = -0.005$
   $\quad -0.005$ is larger

5. Four hundred seventy and seventy-seven
   thousandths = 470.077

6. $\dfrac{4}{9} = 4 \div 9 = 0.\overline{4}$

7. 9,004.0403 Nine thousand, four and
   four hundred three ten-thousandths.

8. 708.056 to the nearest hundred
   700

9. $0.875 = \dfrac{875}{1000} = \dfrac{7}{8}$

10. 939.909 to the nearest tenth
    939.9

11. $\dfrac{3}{7} = 3 \div 7 = 0.429$

12. $\begin{array}{r} 49.930 \\ 604.00 \\ + \ \underline{0.793} \\ 654.723 \end{array}$

13. $(0.325)(76.6)$ $\qquad 35 \times 766 = 26810$
    $= 26.81$

14. $\begin{array}{r} -26.6 \\ +\underline{-45.3} \\ -71.9 \end{array}$ $\qquad \begin{array}{r} 349.004 \\ - \ \underline{71.900} \\ 277.104 \end{array}$

15. $1106.46 \div (-2.7) = -409.8$

$$\begin{array}{r} 409.8 \\ -2.7\overline{)1106.46} \\ \underline{108} \\ 264 \\ \underline{243} \\ 216 \\ 216 \end{array}$$

16. $(-535.5)(9.2)$
    $5355 \times 9.2 = 492660 = -4926.6$

17.
$$\begin{array}{r} 763.870 \\ -\ \underline{84.392} \\ 679.498 \end{array}$$

18. $264.8 \div 1.7 = 155.76$

$$\begin{array}{r} 155.764 \\ 1.7\overline{\smash{)}264.8\,000} \\ \underline{17} \\ 94 \\ \underline{85} \\ 130 \\ \underline{119} \\ 110 \\ \underline{102} \\ 80 \end{array}$$

19.
$$\begin{array}{r} -36.08 \\ +\ \underline{-21.39} \\ -57.47 \end{array}$$

20. $-4.38 - (-7.89) = -4.38 + 7.89$

$$\begin{array}{r} 7.89 \\ -\ \underline{4.38} \\ 3.51 \end{array}$$

21. $8 - 3.3 - 0.5 - 0.45 =$
$8 + (-3.3) + (-0.5) + (-0.45) = 3.75$

22. $4.8 - 0.83 - 9.95 =$
$4.8 + (-0.83) + (-9.95) = -5.98$

23. $0.2 + \dfrac{4}{5} + \dfrac{7}{5} =$

$\dfrac{1}{5} + \dfrac{4}{5} + \dfrac{7}{5} = \dfrac{12}{5} = 2\dfrac{2}{5}$

24. $0.7(\dfrac{3}{8} + 0.125) =$

$0.7(0.375 + 0.125) =$

$0.7(0.5) = 0.35$

25. $(\dfrac{1}{3})^2 (5.4) + (\dfrac{1}{2})^3 (3.2) =$

$(\dfrac{1}{\cancel{3}} \cdot \dfrac{1}{\cancel{3}} \cdot \dfrac{\cancel{54}^{\,3}}{\cancel{10}_{\,5}}) + (\dfrac{1}{\cancel{2}} \cdot \dfrac{1}{\cancel{2}} \cdot \dfrac{1}{\cancel{2}} \cdot \dfrac{\cancel{32}^{\,2}}{\cancel{10}_{\,5}}) =$

$\dfrac{3}{5} + \dfrac{2}{5} = \dfrac{5}{5} = 1$

26. $0.00000765 = 7,65 \times 10^{-6}$

27. $2.48 \times 10^6 = 2,480,000$

28. A dozen is 12
12 cost $23.64
1 costs $23.64 \div 12 = 1.97$
$1.97 for each pen.

29. 1st minute 22 cents
16 minutes at 19 cents each
$0.22 + [16(0.19)] = 0.22 + 3.04 = 3.26$
The call costs $3.26.

30.
Math book	42.97
English book	23.45
History book	2.99
History book	13.54
History book	10.02
Total	92.97

The books will cost $92.97.

CHAPTER 8: EQUATIONS (REVISITED) AND INEQUALITIES

8.1 Solving Equations Using Both Properties

1.
$$21 + a - 2 = \frac{36}{6}$$
$$19 + a = 6$$
$$-19 \quad = -19$$
$$a = -13$$

Check:
$$21 + (-13) - 2 = \frac{36}{6}$$
$$6 = 6$$

2.
$$16 + b - 3 = \frac{24}{8}$$
$$13 + b = 3$$
$$-13 \quad = -13$$
$$b = -10$$

Check:
$$16 + b - 3 = \frac{24}{8}$$
$$16 + (-10) - 3 = 3$$
$$3 = 3$$

3.
$$30 + c - 15 = 10 \cdot 2$$
$$15 + c = 20$$
$$-15 \quad = -15$$
$$c = 5$$

Check:
$$30 + 5 - 15 = 20$$
$$20 = 20$$

4.
$$14 + f - 3 = 11 \cdot 3$$
$$11 + f = 33$$
$$-11 \quad = -11$$
$$f = 22$$

Check:
$$14 + 22 - 3 = 33$$
$$33 = 33$$

5.
$$11z + 2 - 10z = 41$$
$$2 + z = 41$$
$$-2 \quad = -2$$
$$z = 39$$

Check:
$$11(39) + 2 - 10(39) = 41$$
$$429 + 2 - 390 = 41$$
$$41 = 41$$

6.
$$7q + 8 - 6q = 29 + 13$$
$$8 + q = 42$$
$$-8 \quad = -8$$
$$q = 34$$

Check:
$$7(34) + 8 - 6(34) = 29 + 13$$
$$238 + 8 - 204 = 42$$
$$42 = 42$$

7.
$$20x + 17 = 19x - 5$$
$$-19x \quad = -19x$$
$$x + 17 = -5$$
$$-17 = -17$$
$$x = -22$$

Check:
$$20(-22) + 17 = 19(-22) - 5$$
$$-440 + 17 = -418 - 5$$
$$-423 = -423$$

98

8.
$$8b - 10 = 6 + 7b$$
$$\underline{-7b \qquad = -7b}$$
$$b - 10 = 6$$
$$\underline{10 = 10}$$
$$b = 16$$

Check: $8(16) - 10 = 6 + 7(16)$
$$128 - 10 = 6 + 112$$
$$118 = 118$$

9.
$$-5y = 60$$
$$\frac{-5y}{-5} = \frac{60}{-5}$$
$$y = -12$$

Check:
$$-5(-12) = 60$$
$$60 = 60$$

10.
$$-4z = 108$$
$$\frac{-4z}{-4} = \frac{108}{-4}$$
$$z = -27$$

Check:
$$-4(-27) = 108$$
$$108 = 108$$

11.
$$8x = -24$$
$$\frac{8x}{8} = \frac{-24}{8}$$
$$x = -3$$

Check:
$$8(-3) = -24$$
$$-24 = -24$$

12.
$$12w = -120$$
$$\frac{12w}{12} = \frac{-120}{12}$$
$$w = -10$$

Check:
$$12(-10) = -120$$
$$-120 = -120$$

13.
$$2x + 5 = 17$$
$$\underline{-5 = -5}$$
$$\frac{2x}{2} = \frac{12}{2}$$
$$x = 6$$

Check: $2(6) + 5 = 17$
$$12 + 5 = 17$$
$$17 = 17$$

14.
$$2x + 3 = 15$$
$$\underline{-3 = -3}$$
$$\frac{2x}{2} = \frac{12}{2}$$
$$x = 6$$

Check: $2(6) + 3 = 15$
$$12 + 3 = 15$$
$$15 = 15$$

15.
$$2y - 7 = 9$$
$$\underline{7 = 7}$$
$$\frac{2y}{2} = \frac{16}{2}$$
$$y = 8$$

Check: $2(8) - 7 = 9$
$$16 - 7 = 9$$
$$9 = 9$$

16.
$$3y - 8 = 7$$
$$\underline{8 = 8}$$
$$\frac{3y}{3} = \frac{15}{3}$$
$$y = 5$$

Check: $3(5) - 8 = 7$
$$15 - 8 = 7$$
$$7 = 7$$

17.
$$10x + 7 = -3$$
$$-7 = -7$$
$$\frac{10x}{10} = \frac{-10}{10}$$
$$x = -1$$

Check: $10(-1) + 7 = -3$
$-10 + 7 = -3$
$-3 = -3$

18.
$$9x + 8 = -10$$
$$-8 = -8$$
$$\frac{9x}{9} = \frac{-18}{9}$$
$$x = -2$$

Check: $9(-2) + 8 = -10$
$-18 + 8 = -10$
$-10 = -10$

19.
$$15 = 4x + 3$$
$$-3 = -3$$
$$\frac{12}{4} = \frac{4x}{4}$$
$$3 = x$$

Check: $15 = 4(3) + 3$
$15 = 12 + 3$
$15 = 15$

20.
$$17 = 5x + 2$$
$$-2 = -2$$
$$\frac{15}{5} = \frac{5x}{5}$$
$$3 = x$$

Check: $17 = 5(3) + 2$
$17 = 15 + 2$
$17 = 17$

21.
$$-12 = -6x + 6$$
$$-6 = -6$$
$$\frac{-18}{-6} = \frac{-6x}{-6}$$
$$3 = x$$

Check: $-12 = -6(3) + 6$
$-12 = -18 + 6$
$-12 = -12$

22.
$$-14 = -8x + 2$$
$$-2 = -2$$
$$\frac{-16}{-8} = \frac{-8}{-8}$$
$$2 = x$$

Check: $-14 = -8(2) + 2$
$-14 = -16 + 2$
$-14 = -14$

23.
$$7y - 8 = 5y - 16$$
$$-5y =$$
$$2y - 8 = -16$$
$$\frac{2y}{2} = \frac{-8}{2}$$
$$y = -4$$

Check: $7(-4) - 8 = 5(-4) - 16$
$-28 - 8 = -20 - 16$
$-36 = -36$

24.
$$5x + 7 = 3x - 5$$
$$-3x = -3x$$
$$2x + 7 = -5$$
$$-7 = -7$$
$$\frac{2x}{2} = \frac{-12}{2}$$
$$x = -6$$

Check: $5(-6) + 7 = 3(-6) - 5$
$-30 + 7 = -18 - 5$
$-23 = -23$

25.

$$15z + 3 = 5z - 8$$
$$-5z = -5z$$
$$10z + 3 = -8$$
$$ -3 = -3$$
$$\frac{10z}{10} = \frac{-11}{10}$$
$$z = -\frac{11}{10}$$

Check: $\quad 15(-\frac{11}{10}) + 3 = 5(-\frac{11}{10}) - 8$
$$-13\frac{1}{2} = -13\frac{1}{2}$$

26.

$$20s + 5 = 13s - 6$$
$$-13s = -13s$$
$$7s + 5 = -6$$
$$ -5 = -5$$
$$\frac{7s}{7} = \frac{-11}{7}$$
$$s = -\frac{11}{7}$$

Check: $\quad 20(-\frac{11}{7}) + 5 = 13(-\frac{11}{7}) - 6$
$$-26\frac{3}{7} = -26\frac{3}{7}$$

27.

$$-7t + 2 = 4t - 1$$
$$7t = 7t$$
$$2 = 11t - 1$$
$$1 = 1$$
$$\frac{3}{11} = \frac{11t}{11}$$
$$\frac{3}{11} = t$$

Check: $\quad -7(\frac{3}{11}) + 2 = 4(\frac{3}{11}) - 1$
$$\frac{1}{11} = \frac{1}{11}$$

28.

$$-8t + 3 = 3t - 19$$
$$8t = 8t$$
$$3 = 11t - 19$$
$$19 = 19$$
$$\frac{22}{11} = \frac{11t}{11}$$
$$2 = t$$

Check:
$$-8(2) + 3 = 3(2) - 19$$
$$-13 = -13$$

29.

$$3(2y + 1) = 9$$
$$6y + 3 = 9$$
$$-3 = -3$$
$$\frac{6y}{6} = \frac{6}{6}$$
$$y = 1$$

Check: $\quad 3(2y + 1) = 9$
$$3(2(1) + 1) = 9$$
$$3(2 + 1) = 9$$
$$3(3) = 9$$
$$9 = 9$$

30.

$$4(3y + 2) = 20$$
$$12y + 8 = 20$$
$$-8 = -8$$
$$\frac{12y}{12} = \frac{12}{12}$$
$$y = 1$$

Check:
$$4(3(1) + 2) = 20$$
$$4(3 + 2) = 20$$
$$4(5) = 20$$
$$20 = 20$$

31.

$$-2(10x + 7) = 26$$
$$-20x - 14 = 26$$
$$14 = 14$$
$$\frac{-20x}{-20} = \frac{40}{-20}$$
$$x = -2$$

Check:
$$-2(10(-2) + 7) = 26$$
$$-2(-20 + 7) = 26$$
$$-2(-13) = 26$$
$$26 = 26$$

32.

$$-5(3x + 4) = 5$$
$$-15 - 20 = 5$$
$$20 = 20$$
$$\frac{-15x}{-15} = \frac{25}{-15}$$
$$x = -\frac{5}{3}$$

Check:
$$-5(3(-\frac{5}{3}) + 4) = 5$$
$$-5(-5 + 4) = 5$$
$$-5(-1) = 5$$
$$5 = 5$$

33.

$$2(3y - 4) = 4y + 10$$
$$6y - 8 = 4y + 10$$
$$-4y = -4y$$
$$2y - 8 = 10$$
$$8 = 8$$
$$\frac{2y}{2} = \frac{18}{2}$$
$$y = 9$$

Check:
$$2(3(9) - 4) = 4(9) + 10$$
$$2(27 - 4) = 36 + 10$$
$$2(23) = 46$$
$$46 = 46$$

34.

$$5(y + 2) = 3y + 12$$
$$5y + 10 = 3y + 12$$
$$-3y = -3y$$
$$2y + 10 = 12$$
$$-10 = -10$$
$$\frac{2y}{2} = \frac{2}{2}$$
$$y = 1$$

Check:
$$5(1 + 2) = 3(1) + 12$$
$$5(3) = 3 + 12$$
$$15 = 15$$

35.

$$2(3x + 2) - 12 = 3x - 11$$
$$6x + 4 - 12 = 3x - 11$$
$$6x - 8 = 3x - 11$$
$$-3 = -3x$$
$$3x - 8 = -11$$
$$8 = 8$$
$$\frac{3x}{3} = \frac{-3}{3}$$
$$x = -1$$

Check:
$$2(3(-1) + 2) - 12 = 3(-1) - 11$$
$$2(-3 + 2) - 12 = -3 - 11$$
$$2(-1) - 12 = -14$$
$$-2 - 12 = -14$$
$$-14 = -14$$

36.

$$3(2y + 10) - 8 = 3y + 7$$
$$6y + 30 - 8 = 3y + 7$$
$$6y + 22 = 3y + 7$$
$$\underline{-3y = -3y}$$
$$3y + 22 = 7$$
$$\underline{-22 = -22}$$
$$\frac{3y}{3} = \frac{-15}{3}$$
$$y = -5$$

Check:
$$3(2(-5) + 10) - 8 = 3(-5) + 7$$
$$3(-10 + 10) - 8 = -15 + 7$$
$$3(0) - 8 = -8$$
$$0 - 8 = -8$$
$$-8 = -8$$

37.

$$3(x + 7) = 2(2x + 1)$$
$$3x + 21 = 4x + 2$$
$$\underline{-3x = -3x}$$
$$21 = x + 2$$
$$\underline{-2 = -2}$$
$$19 = x$$

Check:
$$3(19 + 7) = 2(2(19) + 1)$$
$$3(26) = 2(38 + 1)$$
$$78 = 2(39)$$
$$78 = 78$$

38.

$$2(x + 5) = 3(2x + 1)$$
$$2x + 10 = 6x + 3$$
$$\underline{-2x = -2x}$$
$$10 = 4x + 3$$
$$\underline{-3 = -3}$$
$$\frac{7}{4} = \frac{4x}{4}$$
$$\frac{7}{4} = x$$

Check: $2(\frac{2}{5} + 5) = 3(2(\frac{7}{4}) + 1)$

$$\frac{7}{2} + 10 = 3(\frac{7}{2} + 1)$$
$$13\frac{1}{2} = 13\frac{1}{2}$$

39.

$$-3(2q - 4) = 8q - 18$$
$$-6q + 12 = 8q - 18$$
$$\underline{6q = -6q}$$
$$12 = 14q - 18$$
$$18 = 18$$
$$\frac{30}{14} = \frac{14q}{14}$$
$$\frac{15}{7} = q$$

Check:
$$-3(2(\frac{15}{7}) - 4) = 8(\frac{15}{7}) - 18$$
$$-\frac{6}{7} = -\frac{6}{7}$$

40.

$$-2(4r + 7) = 2r - 20$$
$$-8r - r = 2r - 20$$
$$8r = 8r$$
$$-14 = 10r - 20$$
$$20 = 20$$
$$\frac{6}{10} = \frac{10r}{10}$$
$$\frac{3}{5} = r$$

Check:
$$-2(4(\frac{3}{5}) + 7) = 2(\frac{3}{5}) - 20$$
$$-18\frac{4}{5} = -18\frac{4}{5}$$

41.

$$2(3y + 2) - 12 = 3y - 11$$
$$6y + 4 - 12 = 3y - 11$$
$$6y - 8 = 3y - 11$$
$$\underline{-3y = -3y}$$
$$8 = 8$$
$$\frac{3y}{3} = \frac{-3}{3}$$
$$y = -1$$

Check: $2(3(-1) + 2) - 2 = 3(-1) - 11$
$$2(-3 + 2) - 12 = -3 - 11$$
$$2(-1) - 12 = -14$$
$$-2 - 12 = -14$$
$$-14 = -14$$

42.

$$4(2y - 1) + 11 = y - 7$$
$$8y - 4 + 11 = y - 7$$
$$8y + 7 = y - 7$$
$$\underline{-y \qquad = -y}$$
$$7y + 7 = -7$$
$$\underline{-7 = -7}$$
$$\frac{7y}{7} = \frac{-14}{7}$$
$$y = -2$$

Check:
$$4(2(-2) - 1) + 11 = -2 - 7$$
$$4(-4 - 1) + 11 = -9$$
$$4(-5) + 11 = -9$$
$$-20 + 11 = -9$$
$$-9 = -9$$

43.

$$3(2 + 5x) - (1 + 14x) = 6 - 12$$
$$6 + 15x - 1 - 14x = -6$$
$$x + 5 = -6$$
$$\underline{-5 = -5}$$
$$x = -11$$

Check:
$$3(2 + 5(-11)) - (1 + 14(-11)) = 6 - 12$$
$$3(2 + (-55) - (1 + (-154)) = -6$$
$$3(-53) - (-153) = -6$$
$$-159 + 153 = -6$$
$$-6 = -6$$

44.

$$2(3 + 2x) - (7 + 10x) = 5 - 9$$
$$6 + 4x - 7 - 10x = -4$$
$$-6x - 1 = -4$$
$$\underline{1 = 1}$$
$$\frac{-6x}{-6} = \frac{-3}{-6}$$
$$x = \frac{1}{2}$$

Check:
$$2(3 + 2(\tfrac{1}{2})) - (7 + 10(\tfrac{1}{2})) = 5 - 9$$
$$2(4) - (7 + 5) = -4$$
$$8 - 12 = -4$$
$$-4 = -4$$

45.

$$4s + 5s - 3s + 8 + (-5s) = 12 - (-8)$$
$$s + 8 = 20$$
$$\underline{-8 = -8}$$
$$s = 12$$

Check:
$$(12) + 5(12) - 3(120 + 8 + (-5(12)) = 12 - (-8)$$
$$48 + 60 - 36 + 8 + (-60) = 20$$
$$20 = 20$$

46.

$$10q + 7q - 3q + 5 - 12q = 16 - (-4)$$
$$2q + 5 = 20$$
$$\underline{-5 = -5}$$
$$\frac{2q}{2} = \frac{15}{2}$$
$$q = \frac{15}{2}$$

Check:
$$10(\tfrac{15}{2}) + 7(\tfrac{15}{2}) - 3(\tfrac{15}{2}) + 5 - 12(\tfrac{15}{2}) = 16 - (-4)$$
$$75 + 52\tfrac{1}{2} - 22\tfrac{1}{2} + -90 = 20$$
$$20 = 20$$

8.2 Solving Equations Containing Fractions

1.
$$\frac{3}{8}x + 1 = \frac{1}{8}$$
$$8(\frac{3}{8}x) + 8(1) = 8(\frac{1}{8})$$
$$3x + 8 = 1$$
$$-8 = -8$$
$$\frac{3x}{3} = \frac{-7}{3}$$
$$x = \frac{-7}{3}$$

Check:
$$\frac{3}{8}(\frac{-7}{3}) + 1 = \frac{1}{8}$$
$$\frac{1}{8} = \frac{1}{8}$$

2.
$$\frac{2}{5}x + 3 = \frac{1}{5}$$
$$5(\frac{2}{5}x) + 5(3) = 5(\frac{1}{5})$$
$$2x + 15 = 1$$
$$-15 = -15$$
$$\frac{2x}{2} = \frac{-14}{2}$$
$$x = -7$$

Check:
$$\frac{2}{5}(-7) + 3 = \frac{1}{5}$$
$$\frac{-14}{5} + 3 = \frac{1}{5}$$
$$\frac{1}{5} = \frac{1}{5}$$

3.
$$\frac{3}{4}y - 2 = \frac{3}{4}$$
$$4(\frac{3}{4}y) - 4(2) = 4(\frac{3}{4})$$
$$3y - 8 = 3$$
$$8 = 8$$
$$\frac{3y}{3} = \frac{11}{3}$$
$$y = \frac{11}{3}$$

Check:
$$(\frac{3}{4} \cdot \frac{11}{3}) - 2 = \frac{3}{4}$$
$$\frac{3}{4} = \frac{3}{4}$$

4.
$$\frac{5}{6}y - 1 = \frac{1}{6}$$
$$6(\frac{5}{6}y) - 6(1) = 6(\frac{1}{6})$$
$$5y - 6 = 1$$
$$6 = 6$$
$$\frac{5y}{5} = \frac{7}{5}$$
$$y = \frac{7}{5}$$

Check:
$$(\frac{5}{6} \cdot \frac{7}{5}) = \frac{1}{6}$$
$$\frac{1}{6} = \frac{1}{6}$$

5.
$$\frac{1}{3}x + \frac{2}{3} = 5$$
$$3(\frac{1}{3}x) + 3(\frac{2}{3}) = 3(5)$$
$$x + 2 = 15$$
$$-2 = -2$$
$$x = 13$$

Check:
$$\frac{1}{3}(13) + \frac{2}{3} = 5$$
$$5 = 5$$

105

6.
$$\frac{1}{6}y + \frac{5}{6} = 2$$
$$6(\frac{1}{6}y) + 6(\frac{5}{6}) = 6(2)$$
$$y + 5 = 12$$
$$-5 = -5$$
$$y = 7$$

Check: $\frac{1}{6}(7) + \frac{5}{6} = 2$
$$2 = 2$$

7.
$$\frac{2}{9}y - \frac{1}{9} = 3$$
$$9(\frac{2}{9}y - 9(\frac{1}{9}) = 9(3)$$
$$2y - 1 = 27$$
$$1 = 1$$
$$\frac{2y}{2} = \frac{28}{2}$$
$$y = 14$$

Check: $\frac{2}{9}(14) - 9 = 3$
$$3 = 3$$

8.
$$\frac{3}{10}x - \frac{1}{10} = 4$$
$$10(\frac{3}{10}x) - 10(\frac{1}{10}) = 10(4)$$
$$3x - 1 = 40$$
$$1 = 1$$
$$\frac{3x}{3} = \frac{41}{3}$$
$$x = \frac{41}{3}$$

Check: $\frac{3}{10}(\frac{41}{3}) - \frac{1}{10} = 4$
$$4 = 4$$

9.
$$4x + \frac{1}{2} = \frac{1}{4}$$
$$4(4x) + 4(\frac{1}{2}) = 4(\frac{1}{4})$$
$$16x + 2 = 1$$
$$-2 = -2$$
$$\frac{16x}{16} = \frac{-1}{16}$$
$$x = -\frac{1}{16}$$

Check:
$$4(-\frac{1}{16}) + \frac{1}{2} = \frac{1}{4}$$
$$\frac{1}{4} = \frac{1}{4}$$

10.
$$5y + \frac{1}{5} = \frac{1}{10}$$
$$10(5y) + 10(\frac{1}{5}) = 10(\frac{1}{10})$$
$$50y + 2 = 1$$
$$-2 = -2$$
$$\frac{50y}{50} = \frac{-1}{50}$$
$$y = -\frac{1}{50}$$

Check:
$$5(-\frac{1}{5}) + \frac{1}{5} = \frac{1}{10}$$
$$\frac{1}{10} = \frac{1}{10}$$

11.

$$3x - \frac{1}{3} = \frac{1}{6}$$

$$6(3x) - 6(\frac{1}{3}) = 6(\frac{1}{6})$$

$$18x - 2 = 1$$

$$2 = 2$$

$$\frac{18x}{18} = \frac{3}{18}$$

$$x = \frac{1}{6}$$

Check:

$$3(\frac{1}{6}) - \frac{1}{3} = \frac{1}{6}$$

$$\frac{1}{6} = \frac{1}{6}$$

12.

$$6x - \frac{1}{8} = \frac{1}{16}$$

$$16(6x) - 16\frac{1}{8} = 16\frac{1}{16}$$

$$96x - 2 = 1$$

$$2 = 2$$

$$\frac{96x}{96} = \frac{3}{96}$$

$$x = \frac{1}{32}$$

check:

$$6(\frac{1}{32}) - \frac{1}{8} = \frac{1}{16}$$

$$\frac{1}{16} = \frac{1}{16}$$

13.

$$\frac{y}{3} + \frac{1}{2} = -\frac{1}{2}$$

$$6(\frac{y}{3}) + 6(\frac{1}{2}) = 6(-\frac{1}{2})$$

$$2y + 3 = -3$$

$$-3 = -3$$

$$\frac{2y}{2} = \frac{-6}{2}$$

$$y = -3$$

Check:

$$\frac{-3}{3} + \frac{1}{2} = -\frac{1}{2}$$

$$-\frac{1}{2} = -\frac{1}{2}$$

14.

$$\frac{x}{2} + \frac{4}{3} = -\frac{2}{3}$$

$$6(\frac{x}{2}) + 6(\frac{4}{3}) = 6(-\frac{2}{3})$$

$$3x + 8 = -4$$

$$-8 = -8$$

$$\frac{3x}{3} = \frac{-12}{3}$$

$$x = -4$$

Check:

$$\frac{-4}{2} + \frac{4}{3} = -\frac{2}{3}$$

$$-\frac{2}{3} = -\frac{2}{3}$$

15.

$$\frac{x}{5} + x = 4$$

$$5(\frac{x}{5}) + 5(x) = 5(4)$$

$$x + 5x = 20$$

$$\frac{6x}{6} = \frac{20}{6}$$

$$x = \frac{10}{3}$$

Check: $\frac{10}{3}(\frac{1}{5}) + \frac{10}{3} = 4$

$$4 = 4$$

16.

$$\frac{x}{3} + x = 8$$

$$3(\frac{x}{3}) + 3(x) = 3(8)$$

$$x + 3x = 24$$

$$\frac{4x}{4} = \frac{24}{4}$$

$$x = 6$$

Check: $\quad \frac{6}{3} + 6 = 8$

$$8 = 8$$

17.

$$\frac{x}{3} + \frac{x}{6} = 4$$

$$6(\frac{x}{3}) + 6(\frac{x}{6}) = 6(4)$$

$$2x + x = 24$$

$$\frac{3x}{3} = \frac{24}{3}$$

$$x = 8$$

Check: $\quad \frac{8}{3} + \frac{8}{6} = 4$

$$4 = 4$$

18.

$$\frac{x}{2} + \frac{x}{5} = 7$$

$$10(\frac{x}{2}) + 10(\frac{x}{5}) = 10(7)$$

$$5x + 2x = 70$$

$$\frac{7x}{7} = \frac{70}{7}$$

$$x = 10$$

Check: $\quad \frac{10}{2} + \frac{10}{5} = 7$

$$7 = 7$$

19.

$$\frac{w}{7} - \frac{w}{2} = 1$$

$$14(\frac{w}{7}) - 14(\frac{w}{2}) = 14(1)$$

$$2w - 7w = 14$$

$$\frac{-5w}{-5} = \frac{14}{-5}$$

$$w = -\frac{14}{5}$$

Check:

$$-\frac{24}{5}(\frac{1}{7}) - (-\frac{14}{5}(\frac{1}{2}) = 1$$

$$-\frac{2}{5} - (-\frac{7}{5} = 1$$

$$-\frac{2}{5} + \frac{7}{5} = 1$$

$$1 = 1$$

20.

$$\frac{v}{4} - \frac{v}{6} = 2$$

$$12(\frac{v}{4}) - 12(\frac{v}{6}) = 12(2)$$

$$3v - 2v = 24$$

$$v = 24$$

Check:

$$\frac{24}{4} - \frac{24}{6} = 2$$

$$6 - 4 = 2$$

$$2 = 2$$

21.

$$\frac{3x}{5} + \frac{7x}{6} = \frac{2}{3}$$

$$30\left(\frac{3x}{5}\right) + 30\left(\frac{7x}{6}\right) = 30\left(\frac{2}{3}\right)$$

$$18x + 35x = 20$$

$$\frac{53x}{53} = \frac{20}{53}$$

$$x = \frac{20}{53}$$

Check:

$$\frac{3}{5}\left(\frac{20}{53}\right) + \frac{7}{6}\left(\frac{20}{53}\right) = \frac{2}{3}$$

$$\frac{12}{53} + \frac{70}{159} = \frac{2}{3}$$

$$\frac{2}{3} = \frac{2}{3}$$

22.

$$\frac{2x}{3} + \frac{3x}{4} = \frac{6}{5}$$

$$60\left(\frac{2x}{3}\right) + 60\left(\frac{3x}{4}\right) = 60\left(\frac{6}{5}\right)$$

$$40x + 45x = 72$$

$$\frac{85x}{85} = \frac{72}{85}$$

$$x = \frac{72}{85}$$

Check:

$$\frac{2}{3}\left(\frac{72}{85}\right) + \frac{3}{4}\left(\frac{72}{85}\right) = \frac{6}{5}$$

$$\frac{48}{85} + \frac{54}{85} = \frac{6}{5}$$

$$\frac{6}{5} = \frac{6}{5}$$

23.

$$\frac{2y}{7} - \frac{3y}{5} = 1$$

$$35\left(\frac{2y}{7}\right) - 35\left(\frac{3y}{5}\right) = 35(1)$$

$$10y - 21y = 35$$

$$\frac{-11y}{-11} = \frac{35}{-11}$$

$$y = -\frac{35}{11}$$

Check:

$$\frac{2}{7}\left(-\frac{35}{11}\right) - \frac{3}{5}\left(-\frac{35}{11}\right) = 1$$

$$-\frac{10}{11} - \left(-\frac{21}{11}\right) = 1$$

$$1 = 1$$

24.

$$\frac{5y}{8} - \frac{2y}{3} = 4$$

$$24\left(\frac{5y}{8}\right) - 24\left(\frac{2y}{3}\right) = 24(4)$$

$$15y - 16y = 96$$

$$-y = 96$$

$$y = -96$$

Check:

$$\frac{5}{8}(-96) - \frac{2}{3}(-96) = 4$$

$$-60 - (-64) = 4$$

$$-60 + 64 = 4$$

$$4 = 4$$

25.

$$\frac{3y}{5} - \frac{1}{2} = \frac{y}{6}$$

$$30(\frac{3y}{5}) - 30(\frac{1}{2}) = 30(\frac{y}{6})$$

$$18y - 15 = 5y$$
$$-18y \qquad = -18y$$
$$\frac{-15}{-13} = \frac{-13y}{-13}$$
$$\frac{15}{13} = y$$

Check:

$$\frac{3}{5}(\frac{15}{13}) - \frac{1}{2} = \frac{1}{6}(\frac{15}{13})$$

$$\frac{9}{13} - \frac{1}{2} = \frac{5}{26}$$

$$\frac{5}{26} = \frac{5}{26}$$

26.

$$\frac{4y}{3} - \frac{1}{4} = \frac{y}{6}$$

$$12(\frac{4y}{3}) - 12(\frac{1}{4}) = 12(\frac{y}{6})$$

$$16y - 3 = 2y$$
$$-16y \qquad = -16y$$
$$\frac{-3}{-14} = \frac{-14y}{-14}$$
$$\frac{3}{14} = y$$

Check:

$$\frac{4}{3}(\frac{3}{14}) - \frac{1}{4} = \frac{1}{6}(\frac{3}{14})$$

$$\frac{2}{7} - \frac{1}{4} = \frac{1}{28}$$

$$\frac{1}{28} = \frac{1}{28}$$

27.

$$\frac{x}{2} - 3 = 7 - \frac{x}{3}$$

$$6(\frac{x}{2}) - 6(3) = 6(7) - 6(\frac{x}{3})$$

$$3x - 18 = 42 - 2x$$
$$2x \qquad = \qquad 2x$$
$$5x - 18 = 42$$
$$18 = 18$$
$$\frac{5x}{5} = \frac{60}{5}$$
$$x = 12$$

Check:

$$\frac{12}{2} - 3 = 7 - \frac{12}{3}$$

$$3 = 3$$

28.

$$\frac{x}{4} - 2 = 5 - \frac{x}{5}$$

$$20(\frac{x}{4}) - 20(2) = 20(5) - 20(\frac{x}{5})$$

$$5x - 40 = 100 - 4x$$
$$4x \qquad = \qquad 4x$$
$$9x - 40 = 100$$
$$40 = 40$$
$$\frac{9x}{9} = \frac{140}{9}$$
$$x = \frac{140}{9}$$

Check:

$$\frac{1}{4}(\frac{140}{9}) - 2 = 5 - \frac{1}{5}(\frac{140}{9})$$

$$\frac{17}{9} = \frac{17}{9}$$

29.

$$\frac{1}{5}(y - 5) = 3$$

$$\frac{1}{2}y - \frac{5}{2} = 3$$

$$2(\frac{1}{2}y) - 2(\frac{5}{2}) = 3(2)$$

$$y - 5 = 6$$
$$5 = 5$$
$$y = 11$$

Check: $\frac{1}{2}(11 - 5) = 3$

$$\frac{1}{2}(6) = 3$$

$$3 = 3$$

30.

$$\frac{1}{3}(y - 2) = 4$$

$$\frac{1}{3}y - \frac{2}{3} = 4$$

$$3(\frac{1}{3}y) - 3(\frac{2}{3}) = 3(4)$$

$$y - 2 = 12$$
$$2 = 2$$
$$y = 14$$

Check: $\frac{1}{3}(14 - 2) = 4$

$$\frac{1}{3}(12) = 4$$

$$4 = 4$$

31.

$$\frac{1}{3}(v + 8) - \frac{1}{4}(3 - 2v) = \frac{1}{6}$$

$$\frac{1}{3}v + \frac{8}{3} - \frac{3}{4} + \frac{1}{2v} = \frac{1}{6}$$

$$12(\frac{1}{3}v) + 12(\frac{8}{3}) - 12(\frac{3}{4}) + 12(\frac{1}{2}v) = 12(\frac{1}{6})$$

$$4v + 32 - 9 + 6v = 2$$
$$10v + 23 = -23$$
$$-23 = -12$$
$$\frac{10v}{10} = \frac{-21}{10}$$
$$v = -\frac{21}{10}$$

Check:

$$\frac{1}{3}(-\frac{21}{10} + 8) - \frac{1}{4}(3 - 2(-\frac{21}{10}) = \frac{1}{6}$$

$$\frac{59}{30} - \frac{9}{5} = \frac{1}{6}$$

$$\frac{1}{6} = \frac{1}{6}$$

32.

$$\frac{1}{3}(6w - 9) = \frac{1}{2}(8w - 4)$$

$$2w - 3 = 4w - 2$$
$$-2w = -2w$$
$$-3 = 2w - 2$$
$$\frac{-1}{2} = \frac{2w}{2}$$
$$-\frac{1}{2} = w$$

Check:

$$\frac{1}{3}(6(-\frac{1}{2}) - 9) = \frac{1}{2}(8(-\frac{1}{2}) - 4)$$

$$\frac{1}{3}(-3 - 9) = \frac{1}{2}(-4 - 4)$$

$$\frac{1}{3}(-12) = \frac{1}{2}(-8)$$

$$-4 = -4$$

111

33.

$$1\frac{1}{2}x + 2 = 3$$
$$\frac{3}{2}x + 2 = 3$$
$$\phantom{\frac{3}{2}x}\ -2 = -2$$
$$\frac{2}{3}\cdot\frac{3}{2}x = 1\cdot\frac{2}{3}$$
$$x = \frac{2}{3}$$

Check:
$$\frac{2}{3}\left(\frac{3}{2}\right) + 2 = 3$$
$$1 + 2 = 3$$
$$3 = 3$$

34.

$$2\frac{1}{3}x + 1 = 4$$
$$\frac{2}{3}x + 1 = 4$$
$$\phantom{\frac{2}{3}x}\ -1 = -1$$
$$\frac{3}{7}\cdot\frac{7}{3}x = 3\cdot\frac{3}{7}$$
$$x = \frac{9}{7}$$

Check:
$$\frac{7}{3}\left(\frac{9}{7}\right) + 1 = 4$$
$$3 + 1 = 4$$
$$4 = 4$$

35.

$$2\frac{1}{4}y - 6 = 8$$
$$\frac{9}{4}y - 6 = 8$$
$$\phantom{\frac{9}{4}y}\ 6 = 6$$
$$\frac{4}{9}\cdot\frac{9}{4}y = 14\cdot\frac{4}{9}$$
$$y = \frac{56}{9}$$

Check:
$$\frac{9}{4}\left(\frac{56}{9}\right) - 6 = 8$$
$$14 - 6 = 8$$
$$8 = 8$$

36.

$$3\frac{1}{5}y - 2 = 3$$
$$\frac{16}{5}y - 2 = 3$$
$$\phantom{\frac{16}{5}y}\ 2 = 2$$
$$\frac{5}{16}\cdot\frac{16}{5}y = 5\cdot\frac{5}{16}$$
$$y = \frac{25}{16}$$

Check:
$$\frac{16}{5}\left(\frac{25}{16}\right) - 2 = 3$$
$$5 - 2 = 3$$
$$3 = 3$$

37.

$$\frac{7}{6}x + \frac{5}{2} = \frac{10}{3}$$

$$6(\frac{7}{6}x) + 5(\frac{5}{2}) = 6(\frac{10}{3})$$

$$7x + 15 = 20$$

$$-15 = -15$$

$$\frac{7x}{7} = \frac{5}{7}$$

$$x = \frac{5}{7}$$

Check:

$$\frac{7}{6}(\frac{5}{7}) + \frac{5}{2} = \frac{10}{3}$$

$$\frac{5}{6} + \frac{5}{2} = \frac{10}{3}$$

$$\frac{10}{3} = \frac{10}{3}$$

38.

$$\frac{9}{8}x + \frac{9}{4} = \frac{7}{2}$$

$$8(\frac{9}{8}x) + 8(\frac{9}{4}) = 8(\frac{7}{2})$$

$$9x + 18 = 28$$

$$-18 = -18$$

$$\frac{9}{9}x = \frac{10}{9}$$

$$x = \frac{10}{9}$$

Check:

$$\frac{9}{8}(\frac{10}{9}) + \frac{9}{4} = \frac{7}{2}$$

$$\frac{5}{4} + \frac{9}{4} = \frac{7}{2}$$

$$\frac{7}{2} = \frac{7}{2}$$

39.

$$\frac{17}{8}y - \frac{7}{6} = \frac{9}{2}$$

$$24(\frac{17}{8}y) - 24(\frac{7}{6}) = 24(\frac{9}{2})$$

$$51y - 28 = 108$$

$$28 = 28$$

$$\frac{51y}{51} = \frac{136}{51}$$

$$y = \frac{8}{3}$$

Check:

$$\frac{17}{8}(\frac{8}{3}) - \frac{7}{6} = \frac{9}{2}$$

$$\frac{17}{3} - \frac{7}{6} = \frac{9}{2}$$

$$\frac{9}{2} = \frac{9}{2}$$

40.

$$\frac{9}{4}y - \frac{3}{2} = \frac{25}{8}$$

$$8(\frac{9}{4}y) - 8(\frac{3}{2}) = 8(\frac{25}{8})$$

$$18y - 12 = 25$$

$$12 = 12$$

$$\frac{18y}{18} = \frac{37}{18}$$

$$y = \frac{37}{18}$$

Check:

$$\frac{9}{4}(\frac{37}{18}) - \frac{3}{2} = \frac{25}{8}$$

$$\frac{37}{8} - \frac{3}{2} = \frac{25}{8}$$

$$\frac{25}{8} = \frac{25}{8}$$

41.
$$\frac{25}{12}x + \frac{43}{6} = \frac{11}{6}x$$
$$12(\frac{25}{12}x) + 12(\frac{43}{6}) = 12(\frac{11}{6}x)$$
$$25x + 86 = 22x$$
$$-25x \qquad = -25x$$
$$\frac{86}{-3} = \frac{-3x}{-3}$$
$$-\frac{86}{3} = x$$

Check:
$$\frac{25}{12}(-\frac{86}{3}) + \frac{43}{6} = \frac{11}{6}(\frac{86}{-3})$$
$$-\frac{473}{9} = -\frac{473}{9}$$

43.
$$\frac{29}{6}x - \frac{13}{2} = \frac{10}{3}x$$
$$6(\frac{29}{6}x) - 6(\frac{13}{2}) = 6(\frac{10}{3}x)$$
$$29x - 39 = 20x$$
$$-29x \qquad = -29x$$
$$\frac{-39}{-9} = \frac{-9x}{-9}$$
$$\frac{13}{3} = x$$

Check:
$$\frac{29}{6}(\frac{13}{3}) - \frac{13}{2} = \frac{10}{3}(\frac{13}{3})$$
$$\frac{130}{9} = \frac{130}{9}$$

42.
$$\frac{6}{5}y + \frac{43}{8} = \frac{13}{5}y$$
$$40(\frac{6}{5}y) + 40(\frac{43}{8}) = 40(\frac{13}{5}y)$$
$$48y + 215 = 104y$$
$$-48y \qquad = -48y$$
$$\frac{215}{56} = \frac{56y}{56}$$
$$\frac{215}{56} = y$$

Check:
$$\frac{6}{5}(\frac{215}{56}) + \frac{43}{8} = \frac{13}{5}(\frac{215}{56})$$
$$\frac{559}{56} = \frac{559}{56}$$

44.
$$\frac{21}{8}x - \frac{49}{12} = \frac{5}{4}x$$
$$24(\frac{31}{8}x) - 24(\frac{49}{12}) = 24(\frac{5}{4}x)$$
$$63x - 98 = 30x$$
$$-63x \qquad = -63x$$
$$\frac{-98}{-33} = \frac{-33x}{-33}$$
$$\frac{98}{33} = x$$

Check:
$$\frac{21}{8}(\frac{98}{33}) - \frac{49}{12} = \frac{5}{4}(\frac{98}{33})$$
$$\frac{245}{66} = \frac{245}{66}$$

45.

$$x + \frac{5}{3} = 3x + \frac{49}{4}$$

$$12(x) + 12(\frac{5}{3}) = 12(3x) + 12(\frac{49}{4})$$

$$12x + 20 = 36x + 147$$
$$-12x = -12x$$
$$20 = 24x + 147$$
$$-147 = -147$$
$$\frac{-127}{24} = \frac{24x}{24}$$
$$\frac{-127}{24} = x$$

Check:

$$-\frac{127}{24} + \frac{5}{3} = 3(-\frac{127}{24}) + \frac{49}{4}$$
$$\frac{-29}{8} = \frac{-29}{8}$$

47.

$$q - \frac{21}{10} = 5q + \frac{16}{5}$$

$$10(q) - 10(\frac{21}{10}) = 10(5q) + 10(\frac{16}{5})$$

$$10q - 21 = 50q + 32$$
$$-10q = -10q$$
$$-21 = 40q + 32$$
$$-32 = -32$$
$$\frac{-53}{40} = \frac{40q}{40}$$
$$\frac{-53}{40} = q$$

Check:

$$-\frac{53}{40} - \frac{21}{10} = 5(-\frac{53}{40}) + \frac{16}{5}$$
$$\frac{-137}{40} = \frac{-137}{40}$$

46.

$$y + \frac{9}{4} = 2y + \frac{51}{5}$$

$$20(y) + 20(\frac{9}{4}) = 20(2y) + 20(\frac{51}{5})$$

$$20y + 45 = 40y + 204$$
$$-20 = -20y$$
$$45 = 20y + 204$$
$$-204 = -204$$
$$\frac{-159}{20} = \frac{20y}{20}$$
$$-\frac{159}{20} = y$$

Check:

$$-\frac{159}{20} + \frac{9}{4} = 2(-\frac{159}{20}) + \frac{51}{5}$$
$$-\frac{57}{10} = -\frac{57}{10}$$

48.

$$z - \frac{28}{9} = 4z + \frac{37}{6}$$

$$18(z) - 18(\frac{28}{9}) = 18(4z) + 18(\frac{37}{9})$$

$$18z - 56 = 72z + 111$$
$$-18z = -18z$$
$$-56 = 54z + 111$$
$$-111 = -111$$
$$\frac{-167}{54} = \frac{542}{54}$$
$$-\frac{167}{54} = z$$

Check:

$$-\frac{167}{54} - \frac{28}{9} = 4(-\frac{167}{54}) + \frac{37}{6}$$
$$\frac{-335}{54} = \frac{-335}{54}$$

49.

$$2(\frac{2}{3}x + \frac{5}{4}) = \frac{37}{12}$$

$$\frac{4}{3}x + \frac{5}{2} = \frac{37}{12}$$

$$12(\frac{4}{3}x) + 12(\frac{5}{2}) = 12(\frac{37}{12})$$

$$16x + 30 = 37$$

$$-30 = -30$$

$$\frac{16x}{16} = \frac{7}{16}$$

$$x = \frac{7}{16}$$

Check:

$$2(\frac{2}{3}(\frac{7}{16}) + \frac{5}{4}) = \frac{37}{12}$$

$$\frac{37}{12} = \frac{37}{12}$$

50.

$$3(\frac{3}{4}x + \frac{7}{3}) = \frac{61}{12}$$

$$\frac{9}{4}x + 7 = \frac{61}{12}$$

$$12(\frac{9}{4}x) + 12(7) = 12(\frac{61}{12})$$

$$27x + 84 = 61$$

$$-84 = -84$$

$$\frac{27x}{27} = \frac{-23}{27}$$

$$x = -\frac{23}{27}$$

Check:

$$3(\frac{3}{4}(-\frac{23}{27}) + \frac{7}{3}) = \frac{61}{12}$$

$$\frac{61}{12} = \frac{61}{12}$$

8.3 Solving Equations Containing Decimals

1.
$$6.054 = 0.43 + x$$
$$6054 = 430 + 1000x$$
$$-430 = -430$$
$$5624 = 1000x$$
$$5.624 = x$$

2.
$$0.079 = 0.408 - x$$
$$79 = 308 - 1000x$$
$$-308 = -308$$
$$-229 = -1000x$$
$$0.229 = x$$

3.
$$4.83 = y + 2.107$$
$$4830 = 1000y + 2107$$
$$-2107 = -2107$$
$$2723 = 1000y$$
$$2.723 = y$$

4.
$$-4.83 = y - 2.107$$
$$-4830 = 1000y - 2107$$
$$2107 = 2107$$
$$-2723 = 1000y$$
$$-2.723 = y$$

5.
$$2z + 0.5 = 0.75$$
$$200z + 50 = 75$$
$$-50 = -50$$
$$\frac{200z}{200} = \frac{25}{200}$$
$$z = 0.125$$

6.
$$0.05z + 17 = 35$$
$$5z + 1700 = 3500$$
$$-1700 = -1700$$
$$\frac{5z}{5} = \frac{1800}{5}$$
$$z = 360$$

7.
$$0.5q + 17 = 14$$
$$5q + 170 = 140$$
$$\underline{\quad -170 = -170}$$
$$\frac{5q}{5} = \frac{-30}{5}$$
$$q = -6$$

8.
$$8.2q - 14 = 17.8$$
$$82q - 140 = 178$$
$$\underline{\quad 140 = 140}$$
$$\frac{82q}{82} = \frac{318}{82}$$
$$q = \frac{159}{41} = 3.88$$

9.
$$0.09 = 0.27 + 3s$$
$$9 = 27 + 3005$$
$$\underline{-27 = -27}$$
$$\frac{-18}{300} = \frac{300s}{300}$$
$$-\frac{3}{50} = s$$
$$s = -\frac{3}{50} = -0.06$$

10.
$$0.9s + 1.1 = 0.2$$
$$9s + 11 = 2$$
$$\underline{\quad -11 = -11}$$
$$\frac{9s}{9} = \frac{-9}{9}$$
$$s = -1$$

11.
$$-7.063 = 9.3 - x$$
$$-7063 = 9300 - 1000x$$
$$\underline{-9300 = -9300}$$
$$\frac{-16363}{-1000} = \frac{-1000x}{-1000}$$
$$-16.363 = x$$

12.
$$-3.4x + 9.2 = -7.1$$
$$-34x + 92 = -71$$
$$\underline{\quad -92 = -92}$$
$$\frac{-34x}{-34} = \frac{-163}{-34}$$
$$x = \frac{163}{34} \approx 4.79$$

13.
$$3 + 0.25x = 0.5x$$
$$300 + 25x = 50x$$
$$\underline{\quad -25x = -25x}$$
$$\frac{300}{25} = \frac{25x}{25}$$
$$12 = x$$

14.
$$12 + 0.62x = -0.7x$$
$$1200 + 62x = -70x$$
$$\underline{\quad -62x = -62x}$$
$$\frac{1200}{-132} = \frac{-132x}{-132}$$
$$-\frac{100}{11} = x$$
$$x = -\frac{100}{11} = 9.0\overline{9}$$

15.
$$1.6h - 15.4 = 2.8h + 22.4$$
$$16h - 154 = 28h + 224$$
$$\underline{-16h \quad\quad = -16h}$$
$$-154 = 12h + 224$$
$$\underline{-224 = \quad\quad -224}$$
$$\frac{-378}{12} = \frac{12h}{12}$$
$$h = -\frac{125}{4} = -31.5$$

16.
$$7g + 23 = 8g - 1.7$$
$$70g + 230 = 80g - 17$$
$$\underline{-70g \quad\quad = -70g}$$
$$230 = 10g - 17$$
$$\underline{17 = \quad\quad 17}$$
$$\frac{247}{10} = \frac{10g}{10}$$
$$g = 24.7$$

17.

$$-1.5c - 1 = -0.5c + 9$$
$$-15c - 10 = -5c + 90$$
$$15c \qquad = 15c$$
$$\qquad -10 = 10c + 90$$
$$\qquad -90 = \qquad -90$$
$$\frac{-100}{10} = \frac{10c}{10}$$
$$c = -10$$

18.

$$0.5c + 0.13 = 0.25c + 1$$
$$50c + 13 = 25c - 100$$
$$-25c \qquad = -25c$$
$$25c \quad 13 = 100$$
$$\qquad -13 = -13$$
$$\frac{25c}{25} = \frac{87}{25}$$
$$c = \frac{87}{25} = 3.48$$

19.

$$0.45x + 0.35x - 17 = -0.15x$$
$$45x + 35x - 1700 = -15x$$
$$80x + 1700 = -15x$$
$$-80x \qquad = -80x$$
$$\frac{-1700}{-95} = \frac{-95x}{-95}$$
$$x = \frac{340}{19}$$

20.

$$4.1x + 0.7x - 2.3 = -8.9x$$
$$41x + 7x - -23 = -89x$$
$$48x - 23 = -89x$$
$$-48x \qquad = -48x$$
$$\frac{-23}{-137} = \frac{-137x}{-137}$$
$$x = \frac{23}{137}$$

21.

$$0.35y + 0.65(70 - y) = 70(0.55)$$
$$0.35y + 45.5 - 0.65y = 38.5$$
$$35y + 4550 - 65y = 3850$$
$$-30y + 4550 = 3850$$
$$-4550 = -4550$$
$$\frac{-30y}{-30} = \frac{-700}{-30}$$
$$y = \frac{70}{3} = 23.\overline{3}$$

22.

$$5x + 5(x - 0.04) = 0.45$$
$$5x + 5x - 0.2 = 0.45$$
$$500x + 500x - 20 = 45$$
$$1000x - 20 = 45$$
$$20 = 20$$
$$\frac{1000x}{1000} = \frac{65}{1000}$$
$$x = 0.065$$

23.

$$-x - 4.25 = \frac{34}{8}$$
$$-x - 4.25 = 4.25$$
$$4.25 = 4.25$$
$$-x = 8.5$$
$$x = -8.5$$

24.

$$-2x - 3.75 = \frac{15}{4}$$
$$-2x - 3.75 = 3.75$$
$$3.75 = 3.75$$
$$\frac{-2x}{-2} = \frac{7.5}{-2}$$
$$x = -3.75$$

25.

$$0.9y - \frac{9}{25} = 0.18y$$
$$0.9y - 0.36 = 0.18y$$
$$90y - 36 = 18y$$
$$-90y \qquad = -90y$$
$$\frac{-36}{-72} = \frac{-72y}{-72}$$
$$y = \frac{1}{2}$$

26.

$$5y - 0.5 = \frac{5}{8}$$

$$5y - 0.5 = 0.625$$

$$5000y - 500 = 625$$

$$500 = 500$$

$$\frac{5000y}{5000} = \frac{1125}{5000}$$

$$y = \frac{9}{40} \qquad 0.225$$

27.

$$0.875z + \frac{5}{6} = \frac{25}{12}$$

$$\frac{7}{8}z + \frac{5}{6} = \frac{25}{12}$$

$$\overset{3}{24}(\frac{7}{8}z) + \overset{4}{24}(\frac{5}{6}) = \overset{2}{24}(\frac{25}{12})$$

$$21z + 20 = 50$$

$$-20 = -20$$

$$\frac{21z}{21} = \frac{30}{21}$$

$$z = \frac{10}{7} = 1.43$$

28.

$$0.75z - \frac{1}{2} = 0.252 + \frac{11}{2}$$

$$\frac{3}{4}z - \frac{1}{2} = \frac{1}{4}z + \frac{11}{2}$$

$$\overset{}{4}(\frac{3}{4}z) - \overset{2}{4}(\frac{1}{2}) = \overset{}{4}(\frac{1}{4}z) + \overset{2}{4}(\frac{11}{2})$$

$$3z - 2 = z + 22$$

$$-z = -z$$

$$2z - 2 = 22$$

$$2 = 2$$

$$\frac{2z}{2} = \frac{24}{2}$$

$$z = 12$$

29.

$$-\frac{1}{2} + 1.25q = \frac{7}{8} - 0.5q$$

$$-\frac{1}{2} + \frac{5}{4}q = \frac{7}{8} - \frac{1}{2}q$$

$$\overset{4}{8}(-\frac{1}{2}) + \overset{2}{8}(\frac{5}{4}q) = 8(\frac{7}{8}) - \overset{4}{8}(\frac{1}{2}q)$$

$$-4 + 10q = 7 - 4q$$

$$4q = 4q$$

$$-4 + 14q = \frac{7}{4}$$

$$4 = 4$$

$$\frac{14q}{14} = \frac{11}{14}$$

$$q = \frac{11}{14} = 0.79$$

30.

$$-\frac{3}{4} + 0.024q = \frac{3}{20} - 0.6q$$

$$-0.75 + 0.24q = 0.15 - 60q$$

$$-75 + 24q = 15 - 60q$$

$$60q = 60q$$

$$-75 + 84q = 15$$

$$75 = 75$$

$$\frac{84q}{84} = \frac{90}{84}$$

$$q = \frac{15}{14}$$

31.

$$\frac{1}{3}s + 5 = 0.2$$

$$\frac{1}{3}s + 5 = \frac{1}{5}$$

$$\overset{5}{15}(\frac{1}{3}s) + 15(5) = \overset{3}{15}(\frac{1}{5})$$

$$5s + 75 = 3$$

$$-75 = -75$$

$$\frac{5s}{5} = \frac{-72}{5}$$

$$s = -\frac{72}{5} = -14.4$$

32.

$$\frac{1}{6}s + 8 = 0.3$$

$$\frac{1}{6}s + 8 = \frac{3}{10}$$

$$30\left(\frac{1}{6}s\right) + 30(8) = 30\left(\frac{3}{10}\right)$$

$$5s + 240 = 9$$

$$-240 = -240$$

$$\frac{5s}{5} = \frac{-231}{5}$$

$$2 = -\frac{231}{5} \text{ or } -46.2$$

33.

$$0.9 - \frac{1}{2}r = 0.25r$$

$$0.9 - 0.5r = 0.25r$$

$$90 - 50r = 25r$$

$$50r = 50r$$

$$\frac{90}{75} = \frac{75r}{75}$$

$$r = \frac{6}{5} = 1.2$$

34.

$$2.5a + 0.7 = \frac{3}{4}a$$

$$2.5a + 0.7 = 0.75a$$

$$250a + 70 = 75a$$

$$-250a = -250a$$

$$\frac{70}{-175} = \frac{-175a}{-175}$$

$$a = -\frac{2}{5} \text{ or } -0.4$$

8.4 Solving Inequalities

1. $\{y \mid y = 56\}$

2. $\{z \mid z = 102\}$

3. $\{x \mid x = -17\}$

4. $\{g \mid g = -29\}$

5. $\{r \mid r \le 18\}$

6. $\{s \mid s \ge 21\}$

7. $\{c \mid c > -51\}$

8. $\{d \mid d < -47\}$

9. $\{x \mid x \ge 0\}$

10. $\{y \mid y \le 0\}$

11.

$$x + 7 < 12$$
$$-7 \quad -7$$
$$x < 5$$

12.

$$x - 3 > 1$$
$$3 \quad 3$$
$$x > 4$$

13.

$$x - 8 \ge -7$$
$$8 \quad 8$$
$$x \ge 1$$

14.

$$x + 8 \le -1$$
$$-8 \quad -8$$
$$x \le -9$$

15.

$$2 + x \le 10$$
$$-2 \quad -2$$
$$x \le 8$$

16.

$$5 + x \ge 11$$
$$-5 \quad -5$$
$$x \ge 6$$

17.

$$5x > 60$$
$$\frac{5x}{5} > \frac{60}{5}$$
$$x > 12$$

120

18.
$$4x < 48$$
$$\frac{4x}{4} < \frac{48}{4}$$
$$x < 12$$

19.
$$9x \geq -36$$
$$\frac{9x}{9} \geq \frac{-36}{9}$$
$$x \geq -4$$

20.
$$8x \leq -56$$
$$\frac{8x}{8} \leq \frac{-56}{8}$$
$$x \leq -7$$

21.
$$-2x < 16$$
$$\frac{-2x}{-2} > \frac{16}{-2}$$
$$x > -8$$

22.
$$-3x > 18$$
$$\frac{-3x}{-3} < \frac{18}{-3}$$
$$x < -6$$

23.
$$-14x \geq -28$$
$$\frac{-14x}{-14} \leq \frac{-28}{-14}$$
$$x \leq 2$$

24.
$$-13x \leq -39$$
$$\frac{-13x}{-13} \geq \frac{-39}{-13}$$
$$x \geq 3$$

25.
$$2x - 7 \leq 9$$
$$\;7\quad\;7$$
$$\frac{2x}{2} \leq \frac{16}{2}$$
$$x \leq 8$$

26.
$$3x - 8 \geq 19$$
$$\;8\quad\;8$$
$$\frac{3x}{3} \geq \frac{27}{3}$$
$$x \geq 9$$

27.
$$10x - 7 < -3$$
$$\;-7\quad -7$$
$$\frac{10x}{10} < \frac{-10}{10}$$
$$x < -1$$

28.
$$9x + 8 > -10$$
$$\;-8\quad -8$$
$$\frac{9x}{9} > \frac{-18}{9}$$
$$x > -2$$

29.
$$4x + 3 \geq 15$$
$$\;-3\quad -3$$
$$\frac{4x}{4} \geq \frac{12}{4}$$
$$x \geq 3$$

30.
$$5x + 2 \leq 17$$
$$\;-2\quad -2$$
$$\frac{5x}{5} \leq \frac{15}{5}$$
$$x \leq 3$$

121

31.
$$-6x + 6 \le -12$$
$$\ -6 \quad -6$$
$$-6 \le -18$$
$$\frac{-6x}{-6} \ge \frac{-18}{-6}$$
$$x \ge 3$$

32.
$$-8x + 2 \ge -14$$
$$\ -2 \quad -2$$
$$-8x \ge -16$$
$$\frac{-8x}{-8} \le \frac{-16}{-8}$$
$$x \le 2$$

33.
$$-14 - 20x < 26$$
$$14 14$$
$$-20x < 40$$
$$\frac{-20x}{-20} > \frac{40}{-20}$$
$$x > -2$$

34.
$$-20 - 15x > 5$$
$$20 20$$
$$-15x > 25$$
$$\frac{-15x}{-15} < \frac{25}{-15}$$
$$x < \frac{-5}{3}$$

35.
$$8 + 12x \ge 20$$
$$-8 -8$$
$$\frac{12x}{12} \ge \frac{12}{12}$$
$$x \ge 1$$

36.
$$12 - 5y \le -48$$
$$-12 -12$$
$$-5y \le -60$$
$$\frac{-5y}{-5} \ge \frac{-60}{-5}$$
$$y \ge 12$$

37.
$$-3x - 6 < -6$$
$$\ 6 \quad 6$$
$$-3x < 0$$
$$\frac{-3x}{-3} > \frac{0}{-3}$$
$$x > 0$$

38.
$$-4 - 9 > -9$$
$$\ 9 \quad 9$$
$$-4x > 0$$
$$\frac{-4x}{-4} < \frac{0}{-4}$$
$$x < 0$$

39.
$$-x + 5 \ge 5$$
$$\ -5 \quad -5$$
$$-x \ge 0$$
$$(-1)(-x) \le 0(-1)$$
$$x \le 0$$

40.
$$-x + 7 \le 7$$
$$\ -7 \quad -7$$
$$-x \le 0$$
$$(-1)(-x) \ge 0(-1)$$
$$x \ge 0$$

41.
$$15 > 3x$$
$$\frac{15}{3} > \frac{3x}{3}$$
$$5 > x$$
$$x < 5$$

122

42.
$$21 < 7x$$
$$\frac{21}{7} < \frac{7x}{7}$$
$$3 < x$$
$$x > 3$$

43.
$$9 \le -8x - 7$$
$$ 7 7$$
$$16 \le -8x$$
$$\frac{16}{-8} \ge \frac{-8x}{-8}$$
$$-2 \ge x$$
$$x \le -2$$

44.
$$22 \ge -5x - 3$$
$$ 3 3$$
$$25 \ge -5x$$
$$\frac{25}{-5} \le \frac{-5x}{-5}$$
$$-5 \le x$$
$$x \ge -5$$

REVIEW

1.   See Section 8.1
2.   See Section 8.1
3.   See Section 8.4
4. a.   Multiplcation Property of Equality
    b.   Addition Property of Equality
    c.   Addition Property of Inequalities
5. a.   See Section 8.1
    b.   See Section 8.1
    c.   See Section 8.1
    d.   See Section 8.3
    e.   See Section 8.3
    f.   See Section 8.1
    g.   See Section 8.4
    h.   See Section 8.4

6.
$$3k - 6 = 18 + 2k$$
$$-2k = -2k$$
$$k - 6 = 18$$
$$ 6 = 6$$
$$k = 24$$

Check:
$$3(24) - 6 = 18 + 2(24)$$
$$72 - 6 = 18 + 48$$
$$66 = 66$$

7.
$$2x + 6 - x = 10$$
$$x + 6 = 10$$
$$-6 = -6$$
$$x = 4$$

Check:
$$2(4) + 6 - 4 = 10$$
$$8 + 6 - 4 = 10$$
$$10 = 10$$

8.
$$7y + (-6y) - 4 = 4$$
$$y - 4 = 4$$
$$4 = 4$$
$$y = 8$$

Check:
$$7(8) + (-6)(8) - 4 = 4$$
$$56 + (-48) - 4 = 4$$
$$4 = 4$$

9.
$$-4a + 3 + 5a = 6$$
$$a + 3 = 6$$
$$-3 = -3$$
$$a = 3$$

Check:
$$-4(3) + 3 + 5(3) = 6$$
$$-12 + 3 + 15 = 6$$
$$6 = 6$$

10.
$$10q = 10q + 3$$
$$-10q = -10q$$
$$0 = 3$$
no solution

123

11.

$7 + t = 3 + t + 4$

$7 + t = 7 + t$

any real number

12.

$12 = r + 8$

$\underline{-8 \qquad\quad -8}$

$\quad 4 = r$

13.

$2y = 36$

$\dfrac{2y}{2} = \dfrac{36}{2}$

$y = 18$

14.

$-6x = 102$

$\dfrac{-6x}{-6} = \dfrac{102}{-6}$

$x = -17$

15.

$7t = -28$

$\dfrac{7t}{7} = \dfrac{-28}{7}$

$t = -4$

16.

$\dfrac{1}{4}z = 40$

$\dfrac{4}{1} \cdot \dfrac{1}{4}z = 40 \cdot \dfrac{4}{1}$

$z = 160$

17.

$-\dfrac{2}{3}w = 20$

$(-\dfrac{3}{2})(-\dfrac{2}{3}x) = 20(-\dfrac{3}{2})$

$x = -30$

18.

$\dfrac{2}{5}x = -\dfrac{3}{10}$

$\dfrac{5}{2} \cdot \dfrac{2}{5}x = -\dfrac{3}{10} \cdot \dfrac{5}{2}$

$x = -\dfrac{3}{4}$

19.

$3y = 2\dfrac{1}{3}$

$3y = \dfrac{7}{3}$

$\dfrac{1}{3} \cdot 3y = \dfrac{7}{3} \cdot \dfrac{1}{3}$

$y = \dfrac{7}{9}$

20.

$3y + 8 = -16$

$\qquad\;\, -8 = -8$

$\dfrac{3y}{3} = \dfrac{-24}{3}$

$y = -8$

21.

$-3 = 2x + 7$

$\underline{-7 = \qquad\;\; -7}$

$\dfrac{-10}{2} = \dfrac{2x}{2}$

$-5 = x$

22.

$7 + 5y = 12 + y$

$\underline{\quad\;\; -y = \qquad -y}$

$7 + 4y = 12$

$\underline{-7 \qquad\;\; = -7}$

$\dfrac{4y}{4} = \dfrac{5}{4}$

$y = \dfrac{5}{4}$

23.

$-q + 10 = 2q + 11$

$\;q \qquad\quad\;\; = q$

$\qquad\; 10 = 3q + 11$

$\;-11 = \qquad\; -11$

$\dfrac{-1}{3} = \dfrac{3q}{3}$

$-\dfrac{1}{3} = q$

124

24.
$$3(5h + 2) = 13h + 4$$
$$15h + 6 = 13h + 4$$
$$-13h \quad = -13h$$
$$2h + 6 = 4$$
$$-6 = -6$$
$$\frac{2h}{2} = \frac{-2}{2}$$
$$h = -1$$

25.
$$-2(7 + w) + 20 = -4w + 4$$
$$-14 - 2w + 20 = -4w + 4$$
$$-2w + 6 = -4w + 4$$
$$4w \quad = 4w$$
$$2w + 6 = 4$$
$$-6 = -6$$
$$\frac{2w}{2} = \frac{-2}{2}$$
$$w = -1$$

26.
$$\frac{x}{4} - x = 4$$
$$5(\frac{x}{5}) - 5(x) = 5(4)$$
$$x - -5x = 20$$
$$\frac{-4x}{-4} = \frac{20}{-4}$$
$$x = -5$$

27.
$$3y + \frac{1}{2} = \frac{1}{4}$$
$$4(3y) + 4(\frac{1}{2}) = 4(\frac{1}{4})$$
$$12y + 2 = 1$$
$$-2 = -2$$
$$\frac{12y}{12} = \frac{-1}{12}$$
$$y = -\frac{1}{12}$$

28.
$$\frac{t}{2} + \frac{3}{4} = -\frac{2}{3}$$
$$12(\frac{t}{2}) + 12(\frac{3}{4}) = (-\frac{2}{3}12$$
$$6t + 9 = -8$$
$$-9 = -9$$
$$\frac{6t}{6} = \frac{-17}{6}$$
$$t = \frac{-17}{6}$$

29.
$$\frac{2w}{7} - \frac{w}{2} = \frac{15}{14}$$
$$14(\frac{2w}{7}) - 14(\frac{w}{2}) = 14(\frac{15}{14})$$
$$4w - 7w = 15$$
$$\frac{-3w}{-3} = \frac{15}{-3}$$
$$w = -5$$

30.
$$2\frac{1}{2}q + 5 = 7$$
$$\frac{5}{2}q + 5 = 7$$
$$-5 = -5$$
$$\frac{2}{5} \cdot \frac{5}{2}q = 2 \cdot \frac{2}{5}$$
$$q = \frac{4}{5}$$

31.
$$-1\frac{1}{3}z + \frac{1}{2} = 1$$
$$-\frac{4}{3}z + \frac{1}{2} = 1$$
$$6(-\frac{4}{3}z) + 6(\frac{1}{2}) = 6(1)$$
$$-8z + 3 = 6$$
$$-3 = -3$$
$$\frac{-8z}{-8} = \frac{3}{-8}$$
$$z = -\frac{3}{8}$$

125

32.

$$7x + 2.3 = 8x - 1.7$$
$$70x + 23 = 80x - 17$$
$$\underline{-70x \qquad\quad = -70x}$$
$$23 = 10x - 17$$
$$\underline{17 = \qquad\quad 17}$$
$$\frac{40}{10} = \frac{10x}{10}$$
$$4 = x$$

33.

$$4.2 - x = 8.9$$
$$42 - 10x = 89$$
$$\underline{-42 \qquad\quad = -42}$$
$$\frac{-10x}{-10} = \frac{47}{-10}$$
$$x = -\frac{47}{10} = -4.7$$

34.

$$0.9x - 0.36 = 0.18x$$
$$90x - 36 = 18x$$
$$\underline{-90 \qquad\quad = -90x}$$
$$\frac{-36}{-72} = \frac{-72x}{-72}$$
$$0.5 = \frac{1}{2} = x$$

35.

$$7.2x - 8.37 = 6.03$$
$$720x - 837 = 603$$
$$\underline{\qquad\quad 837 = 837}$$
$$\frac{720x}{720} = \frac{1440}{720}$$
$$x = 2$$

36.

$$1.6x - 15.4 = 2.8x + 22.4$$
$$16x - 154 = 28x + 224$$
$$\underline{-16x \qquad\quad = -16x}$$
$$-154 = 12x + 224$$
$$\underline{-224 = \qquad\quad -224}$$
$$\frac{-378}{12} = \frac{12x}{12}$$
$$x = 31.5 = \frac{63}{2}$$

37.

$$0.3y - 4.2 = (y - 3) + 1.7$$
$$0.3y - 4.2 = y - 3 + 1.7$$
$$3y - 42 = 10y - 30 + 1.7$$
$$3y - 42 = 10y - 13$$
$$\underline{-3y \qquad\quad = -3y}$$
$$-42 = 7y - 13$$
$$\underline{13 = \qquad\quad 13}$$
$$\frac{-29}{7} = \frac{7y}{7}$$
$$-\frac{29}{7} = y$$

38.

$$2x - 3.75 = \frac{15}{4}$$
$$2x - 3.75 = 3.75$$
$$\underline{3.75 = 3.75}$$
$$\frac{2x}{2} = \frac{7.5}{2}$$
$$x = 3.75 = \frac{15}{4}$$

39.

$$-x + 4.5 = \frac{21}{3}$$
$$-x + 4.5 = 7$$
$$\underline{-4.5 = -4.5}$$
$$-x = 2.5$$
$$x = -2.5 = -2\frac{1}{2}$$

40.

$$0.52 + \frac{2}{5} = 0.24y$$
$$0.52 + 0.4 = 0.24y$$
$$52 + 40 = 24y$$
$$\frac{92}{24} = \frac{24y}{24}$$
$$3.8\overline{3} = \frac{23}{6} = y$$

126

41.

$$0.4(0.2 - w) = \frac{3}{10}$$
$$0.08 - 0.4w = 0.3$$
$$8 - 40w = 30$$
$$\underline{-8 \qquad\quad = -8}$$
$$\frac{-40w}{-40} = \frac{22}{-40}$$
$$w = -\frac{11}{20} = -0.55$$

42.

$$0.5n + 0.75n = 3\frac{1}{8}$$
$$0.5n + 0.75n = 3.125$$
$$500n + 750n = 3125$$
$$\frac{1250n}{1250} = \frac{3125}{1250}$$
$$n = \frac{5}{2} = 2.5$$

43.

$$\frac{2}{3}y - 2.5 = \frac{5}{6}$$
$$\frac{2}{3}y - \frac{5}{2} = \frac{5}{6}$$
$$6(\frac{2}{3}y) - 6(\frac{5}{2}) = 6(\frac{5}{6})$$
$$4y - 15 = 5$$
$$15 = 15$$
$$\frac{4y}{4} = \frac{20}{4}$$
$$y = 5$$

44.

$$x + 12 > 8$$
$$\underline{-12 \quad -12}$$
$$x > -4$$

45.

$$7 + x < 3$$
$$\underline{-7 \qquad -7}$$
$$x < -4$$

46.

$$x - 15 \ge -20$$
$$\underline{15 \quad 15}$$
$$x \ge -5$$

47.

$$-3 + x \le -3$$
$$\underline{3 \qquad 3}$$
$$x \le 0$$

48.

$$4x \le 32$$
$$\frac{4x}{4} \le \frac{32}{4}$$
$$x \le 8$$

49.

$$7x > -35$$
$$\frac{7x}{7} > \frac{-35}{7}$$
$$x > -5$$

50.

$$-11x < 44$$
$$\frac{-11x}{-11} > \frac{44}{-11}$$
$$x > -4$$

51.

$$-13x \le -52$$
$$\frac{-13x}{-13} \ge \frac{-52}{-13}$$
$$x \ge 4$$

52.

$$5x + 17 \ge 22$$
$$\underline{-17 \qquad -17}$$
$$\frac{5x}{5} \ge \frac{5}{5}$$
$$x \ge 1$$

53.

$$20 + 7x < -1$$
$$\underline{-20 \qquad\quad -20}$$
$$\frac{7x}{7} < \frac{-21}{7}$$
$$x < -3$$

127

54.

$$-4x + 7 \leq 15$$
$$\underline{\ -7 \quad -7}$$
$$-4x \leq 8$$
$$\frac{-4x}{-4} \geq \frac{8}{-4}$$
$$x \geq -2$$

55.

$$-20 - 6x > 4$$
$$\underline{20 \qquad\qquad 20}$$
$$-6x > 24$$
$$\frac{-6x}{-6} < \frac{24}{-6}$$
$$x < -4$$

56.

$$17 > 2x - 1$$
$$\underline{1 \qquad\qquad 1}$$
$$\frac{18}{2} > \frac{2x}{2}$$
$$9 > x$$
$$x < 9$$

PRACTICE TEST

1. B
2. C
3. A
4. D
5. C
6. C

7.
$$52 = 25$$
$$\frac{52}{2} = \frac{25}{5}$$
$$z = 5$$

8.
$$y + 5 = 20$$
$$\underline{-5 = -5}$$
$$y = 15$$

9.
$$2x + 3 = 7$$
$$\underline{-3 = -3}$$
$$\frac{2x}{2} = \frac{4}{2}$$
$$x = 2$$

10.
$$q + 4 = q$$
$$\underline{-q = -q}$$
$$4 = 0$$
no solution

11.
$$x + 5 > 3$$
$$\underline{-5 \quad -5}$$
$$x > -2$$

12.
$$-14x \leq -56$$
$$\frac{-14x}{-14} \geq \frac{-56}{-14}$$
$$x \geq 4$$

13.
$$-4y - 10 = -2$$
$$\underline{10 = 10}$$
$$\frac{-4y}{-4} = \frac{8}{-4}$$
$$y = -2$$

14.
$$8t = 2\frac{1}{4}$$
$$\frac{1}{8} \cdot 8t = \frac{9}{4} \cdot \frac{1}{8}$$
$$t = \frac{9}{32}$$

15.
$$5a + 4 = 3a + 8$$
$$\underline{-3a = -3a}$$
$$2a + 4 = 8$$
$$\underline{-4 = -4}$$
$$\frac{2a}{2} = \frac{4}{2}$$
$$a = 2$$

128

16.

$$-2 + 2w + 5 = 3w + 3 - w$$
$$2w + 3 = 2w + 3$$
$$\underline{-2w \quad\quad = -2w}$$
$$3 = 3$$

Any real number

17.

$$2b + 15 = -3b + 10$$
$$\underline{3b \quad\quad = 3b}$$
$$56 + 15 = 10$$
$$\underline{-15 = -15}$$
$$\frac{5b}{5} = \frac{-5}{5}$$
$$b = -1$$

18.

$$7 + \frac{c}{5} = 10$$
$$\underline{-7 \quad\quad = -7}$$
$$5(\frac{c}{5}) = 3(5)$$
$$c = 15$$

19.

$$3(s - 5) = 2(s - 3)$$
$$3s - 15 = 2s - 6$$
$$\underline{-2s \quad\quad = -2s}$$
$$s - 15 = -6$$
$$\underline{15 = 15}$$
$$s = 9$$

20.

$$-12 = -x + 2$$
$$\underline{-2 = \quad\quad -2}$$
$$-14 = -x$$
$$x = 14$$

21.

$$0.4(x - 3) + 0.5x = 0.6x$$
$$0.4x - 1.2 + 0.5x = 0.6x$$
$$4x - 12 + 5x = 6x$$
$$9x - 12 = 6x$$
$$\underline{-9x \quad\quad = -9x}$$
$$\frac{-12}{-3} = \frac{-3x}{-3}$$
$$4 = x$$

22.

$$\frac{3a}{4} + 15 = 8$$
$$\underline{-15 = -15}$$
$$\frac{4}{3} \cdot \frac{3a}{4} = -7 \cdot \frac{4}{3}$$
$$a = -\frac{28}{3} = 9.\overline{3}$$

23.

$$\frac{x}{2} - \frac{5}{6} = 12 - \frac{3x}{2}$$
$$6(\frac{x}{2}) - 6(\frac{5}{6}) = 6(12) - 6(\frac{3x}{2})$$
$$3x - 5 = 72 - 9x$$
$$\underline{9x \quad\quad = \quad\quad 9x}$$
$$12x - 5 = 72$$
$$\underline{5 = 5}$$
$$\frac{12x}{12} = \frac{77}{12}$$
$$x = \frac{77}{12} = 6.41\overline{6}$$

24.

$$\frac{x + 2}{4} - \frac{2x + 3}{3} = 12$$
$$12(\frac{x + 2}{4}) - 12(\frac{2x + 3}{3}) = 12.12$$
$$3(x + 2) - 4(2x + 3) = 144$$
$$3x + 6 - 8x - 12 = 144$$
$$-5x - 6 = 144$$
$$\underline{6 = 6}$$
$$\frac{-5x}{-5} = \frac{150}{-5}$$
$$x = -30$$

25.

$$0.3(2 - x) = 0.4(0.7 - x) + 0.3$$
$$0.6 - 0.3x = 0.28 - 0.4x + 0.3$$
$$60 - 30x = 28 - 40x + 30$$
$$60 - 30x = 58 - 40x$$
$$\underline{\quad 40x = \quad\quad 40x}$$
$$60 + 10x = 58$$
$$\underline{-60 \quad\quad = -60}$$
$$\frac{10x}{10} = \frac{-2}{10}$$
$$x = -\frac{1}{5} = -0.2$$

26.

$$\frac{x}{8} - 1 = \frac{3}{4}$$
$$8(\frac{x}{8}) - 8(1) = 8(\frac{3}{4})$$
$$x - 8 = 6$$
$$8 = 8$$
$$x = 14$$

27.

$$\frac{2x}{3} - \frac{3x}{4} = \frac{5}{6}$$
$$12(\frac{2x}{3}) - 12(\frac{3x}{4}) = 12(\frac{5}{6})$$
$$8x - 9x = 10$$
$$-x = 10$$
$$x = -10$$

28.

$$-2x - 3.75 = \frac{15}{4}$$
$$-2x - 3.75 = 3.75$$
$$3.75 = 3.75$$
$$\frac{-2x}{-2} = \frac{7.5}{-2}$$
$$x = -3.75 = -\frac{15}{4}$$

29.

$$5x - 7 \geq -42$$
$$\underline{\quad 7 \quad\quad 7}$$
$$\frac{5x}{5} \geq \frac{-35}{5}$$
$$x \geq -7$$

30.

$$3 - 12x < 39$$
$$\underline{-3 \quad\quad -3}$$
$$-12x < 36$$
$$\frac{-12x}{-12} > \frac{36}{-12}$$
$$x > -3$$

31.

$$4\frac{1}{5} - x = 8.9$$
$$4.2 - x = 8.9$$
$$42 - 10x = 89$$
$$\underline{-42 \quad\quad = -42}$$
$$\frac{-20x}{-10} = \frac{47}{-10}$$
$$x = -\frac{47}{10} = -4.7$$

32.

$$5 - 8v = 21$$
$$\underline{-5 \quad\quad = -5}$$
$$\frac{-8v}{-8} = \frac{16}{-8}$$
$$v = -2$$

33.

$$4c - 3 = 3c + 12$$
$$\underline{-3c \quad\quad = -3c}$$
$$c - 3 = 12$$
$$3 = 3$$
$$c = 15$$

34.

$$0.5n + 4 = 10$$
$$\underline{-4 = -4}$$
$$\frac{0.5n}{0.5} = \frac{6}{0.5}$$
$$n = \frac{60}{5} = 12$$

35.
$$0.8q + 1.2q = -12$$
$$8q + 12q = -120$$
$$\frac{20q}{20} = \frac{-120}{20}$$
$$q = -6$$

36.
$$5[2y - (3y + 1) - y] = -5$$
$$5[2y - 3y - 1 - y] = -5$$
$$10y - 15y - 5 - 5y = -5$$
$$-10y - 5 = -5$$
$$5 = 5$$
$$\frac{-10y}{-10} = \frac{0}{-10}$$
$$y = 0$$

37.
$$15 = 3b - (b - 7)$$
$$15 = 3b - b + 7$$
$$15 = 2b + 7$$
$$-7 = \quad -7$$
$$\frac{8}{2} = \frac{2b}{2}$$
$$4 = b$$

38.
$$\frac{5}{-15} = \frac{-15y}{-15}$$
$$-\frac{1}{3} = y$$

39.
$$-1.3 - 0.7z = 0.42 - 1.1z$$
$$-130 - 70z = 42 - 110z$$
$$110z = \quad 110z$$
$$-130 + 40z = 42$$
$$130 \quad = 130$$
$$\frac{40z}{40} = \frac{172}{40}$$
$$z = 4.3 = \frac{43}{10}$$

40.
$$0.9t - 0.36 = 0.18t$$
$$90t - 36 = 18t$$
$$-90t \quad = -90t$$
$$\frac{-36}{-72} = \frac{-72t}{-72}$$
$$\frac{1}{2} = t$$

41.
$$\frac{1}{3}x + 0.5 = 1\frac{2}{3}$$
$$\frac{1}{3}x + \frac{1}{2} = \frac{5}{3}$$
$$6(\frac{1}{3}x) + 6(\frac{1}{2}) = 6(\frac{5}{3})$$
$$2x + 3 = 10$$
$$-3 = -3$$
$$\frac{2x}{2} = \frac{7}{2}$$
$$x = \frac{7}{2} = 3.5$$

42.
$$0.7y - 0.05 = \frac{1}{5}y + 0.35$$
$$0.7y - 0.05 = 0.02y + 0.35$$
$$70y - 5 = 20y + 35$$
$$-20y \quad = -20y$$
$$50y - 5 = 35$$
$$5 = 5$$
$$\frac{50y}{50} = \frac{40}{50}$$
$$y = \frac{4}{5} = 0.8$$

43.
$$5.4(h + 1) = \frac{1}{2}(1.5 + h)$$
$$5.4h + 5.4 = 0.75 + 0.5h$$
$$540h + 540 = 75 + 50h$$
$$-50 \quad = \quad -50h$$
$$490h + 540 = 75$$
$$-540 = -540$$
$$\frac{490h}{490} = \frac{-465}{490}$$
$$h = -\frac{93}{98} = -0.95$$

44.

$$\frac{1}{4}(1 - k) = 0.7(2 + 3k)$$

$$0.25 - 0.25k = 1.4 + 2.1k$$

$$25 - 25k = 140 + 210K$$

$$25k = 25k$$

$$25 = 140 + 235k$$

$$-140 = -140$$

$$\frac{-115}{235} = \frac{235k}{235}$$

$$k = -\frac{23}{47} = -0.49$$

50.

$$11 \leq 2x + 1$$

$$-1 -1$$

$$\frac{10}{2} \leq \frac{2x}{2}$$

$$5 \leq x$$

$$x \geq 5$$

45.

$$3x \leq 12$$

$$\frac{3x}{3} \leq \frac{12}{3}$$

$$x \leq 4$$

46.

$$-2x > 16$$

$$\frac{-2x}{-2} < \frac{16}{-2}$$

$$x < -8$$

47.

$$x + 7 < 8$$

$$-7 -7$$

$$x < 1$$

48.

$$3x - 1 \geq -3$$

$$1 1$$

$$\frac{3x}{3} \geq \frac{-12}{3}$$

$$x \geq -4$$

49.

$$5 - 2x > 7$$

$$-5 -5$$

$$-2x > 2$$

$$\frac{-2x}{-2} < \frac{2}{-2}$$

$$x < -1$$

CHAPTER 9: PROBLEM SOLVING

9.1 Engaging in Problem Solving

Answers will vary.

9.2 Some Strategies for Solving Logic and Mathematics Problems

1. How much will you pay for 15 magazines at $2.50 each?
   $15(2.5) = 37.5$ $37.50

   The subscription rate for 52 weeks is $22.50.
   $37.50 - 22.50 = 15$

   You will save $15 if you take the subscription rate and you will have the magazine for a year.

2. "Sq fi qz rgs sq fi, shus mo
   "To be or not to be, that is

   djo bwiosmgr."
   the question."

3. $6(8 - 3) + 2$
   $(6 \cdot 2 \div 3) 8$
   $(6 \cdot 8 \div 2) + 8$
   Answers may vary.

4. 2 bins   = 2 bins
   2 bins each contain 2 bins = 4 bins
   4 bins each contain 3 bins = 12 bins
                           Total   = 18 bins

5. A B A C B C B $\underline{D}$ $\underline{C}$ $\underline{D}$

   AB
   AC
   BC
   BD
   CD

6. $5, 2, 4, 1, 2, -1, -2, \underline{-5}$

   $5 - 3 = 2$
   $2 + 2 = 4$
   $4 - 3 = 1$
   $1 + 1 = 2$
   $2 - 3 = -1$
   $-1 + (-1) = -2$
   $-2 - 3 = -5$

7.
	Females	Males	Total
Jason	6	2	8
Rebecca	4	6	10
Cliff	2	10	12

   2 males   Jason
   10 males   Cliff

8. Since $6 = 2 \cdot 3$, both numbers must be divisible by both 2 and 3. This means that they must be even and that the sum of their digits must be divisible by 3.

   The number x2,741,008 has 8 as its last digit, so it is even. The number 1,705,03x has x as its last digit, so x must be an even digit, i.e., $x \in \{0, 2, 4, 6, 8\}$.

   $x + 2 + 7 + 4 + 1 + 0 + 0 + 8 = x + 22$, so $x + 22$ must be divisible by 3.

   $1 + 7 + 0 + 5 + 0 + 3 + x = x + 16$, so $x + 16$ must be divisible by 3.

To see which values of x fulfill these conditions, make a table.

x	x + 22	x + 16
0	22	16
2	**24**	**18**
4	26	20
6	28	22
8	**30**	**24**

Conclusion: Only 2, and 8 result in sums which are dvisible by 3, so $x \in \{2, 8\}$

Check: **22**,741,008 ÷ 6 = 3,790,168 and 1,705,03**2** ÷ 6 = 284,172 and **82**,741,008 ÷ 6 = 13,790,168 and 1,705,03**8** ÷ 6 = 284.173

9.
```
 479        947
+625       +256
1203       1203

 246
+957       Answers may vary.
1203
```

10. A place value table of these base 13 numbers looks like this:

base 13	13^3	13^2	13^1	13^0
a.			7	5
b.		3	Υ	4
c.	4	0	1	0
d.		2	α	π

Remember, their number 10 is 13 in the decimal or base 10 system, so

a. $7 \cdot 13^1 + 5 \cdot 13^0 + 91 + 5 = $ **96**

b. $3 \cdot 13^2 + 10 \cdot 13^1 + 4 \cdot 13^0 = $ $3 \cdot 169 + 10 \cdot 13 + 5 = $ $507 + 130 + 4 = $ **641**

c. $4 \cdot 13^3 + 0 \cdot 13^2 + 1 \cdot 13^1 + 0 \cdot 13^0 = 4 \cdot 2197 + 1 \cdot 13 = 8788 + 13 = $ **8801**

d. $2 \cdot 13^2 + 12 \cdot 13^1 + 11 \cdot 13^0 = $ $2 \cdot 169 + 156 + 11 = 388 + 167 = $ **505**

11.
The man	1
The man going to St. Ives	1
7 wives	7
7 x 7 cats	49
49 x 7 kits	343
343 x 7 mice	2401
Total people and animals	2802

12.
Spaghetti only	82
Spaghetti and pizza	14
Pizza only	140
Chicken only	28
Head count	264

13. Maria is 35, attends Red College, and takes Math 501. Evelyn is 21, attends Olive Community College and takes Engl 700. LaTisha is 19, attends Green State University, and takes Psych 205.

	19	21	35	GSU	OCC	RC	ENGL	MATH	PSYCH
E	x	•	x	x	•	x	•	x	x
L	•	x	x	•	x	x	x	x	•
M	x	x	•	x	x	•	x	•	x

134

9.3 Solving Problems by Using Formulas and Equations

1. Information: number of children expected to be born this year. Number of children who have been born this year.

   Question: How many more births are expected this year?

2. Information: a student's grades on 5 tests.

   Question: What is the student's average?

3. Information: the results of a series of arithmetic operations which are performed on a number.

   Question: What is the number?

4. Information: the price of a pen. Announcement of a half-price sale.

   Question: What is the price of 3 pens?

5. Information: tuition cost for last 20 years.

   Question: How much has the tuition increased?

6. Let m = amouont of money Zak has; 3m = amount of money his brother has.

7. Let y = life expectancy for men in years; y + 6.8 = life expectancy for women in years.

8. Let g = mpg your car gets; g − 5.1 = mpg Chelsey's car gets.

9. Let l = average household income of single parents in 1990.

10. Let r = average driving rate of Connor's parents.

11.
$$
\begin{array}{ll}
4.2 \text{ million} & \text{expected to be born} \\
-\underline{1.7 \text{ million}} & \text{born so far} \\
2.5 \text{ million} & \text{expected to be born}
\end{array}
$$

12. $\dfrac{94 + 82 + 89 + 75 + 79}{5} = \dfrac{419}{5}$

    The student's average is 83.8

13. The number x
$$
\begin{aligned}
2x + 6 &= 24 \\
-6 &= -6 \\
\frac{2x}{2} &= \frac{18}{2} \\
x &= 9
\end{aligned}
$$
    The number is 9.

14. The number x
$$
\begin{aligned}
98 - 7x &= 21 \\
-98 &= -98 \\
\frac{-7x}{-7} &= \frac{-77}{-7} \\
x &= 11
\end{aligned}
$$
    The number is 11.

15. Christian's income x
    his wife's income 2x
$$
\begin{aligned}
x + 2x &= 53{,}135 \\
\frac{3x}{3} &= \frac{53{,}135}{3} \\
x &= 17711.67
\end{aligned}
$$
    Christian earns approx. $17,711.67.

16. Let x = life expectancy for men
    x + 6.8 = lfe expectancy for women
$$
\begin{aligned}
x + 6.8 &= 78.8 \\
-6.8 &= -6.8 \\
x &= 72
\end{aligned}
$$
    72 years was the life expectancy for men.

135

17.
$17,500   income in 1970
$ 3,900   increase in income
$21,400

18.

$$d = r \cdot t$$
$$172 = r(2\tfrac{1}{2})$$
$$\frac{2}{5} \cdot 172 = r \cdot \frac{5}{2} \cdot \frac{2}{5}$$
$$68.8 = r$$

Their average driving rate was 68.8.

19.
One dozen (12) costs $49.50
1 rose costs $49.50 \div 12 = 4.125$
5 roses cost $(4.125)5 = 20.625$

5 roses costs approximately $20.63.

20.
1.2 inches her hour
$(1.2)7$ inches in 7 hours
8.4 inches of snow in 7 hours

21.
$10:10 - 11:10 = 1$ hour
$11:10 - 12:10 = 1$ hour
$12:10 - 12:20 = 10$ minutes $=$

$\frac{1}{6}$ of an hour

Cruise control on for $2\tfrac{1}{6}$ hours
$$R \cdot T = D$$
$$65(2\tfrac{1}{6}) = D$$
$$65 \cdot \frac{13}{6} = D$$
$$140.83 = D$$
140.83 miles driven

22.
$$P = 2l + 2w$$
$$56 = 2(3x) + 2(x)$$
$$56 = 6x + 2x$$
$$\frac{56}{8} = \frac{8x}{8}$$
$$7 = x$$

The width is 7 meters, the length is 21 meters.

23.
12 inch pizza $8.40
8 inch pizza $4.20
16 inch pizza $16.80

2 - 12 inch pizzas	$16.80
1 - 8 inch pizza	$ 4.20
3 - 16 inch pizzas	$50.40
Total	$71.40

The pizzas cost $71.40

24.
Student's GPA is $x + 0.25$
His brother's GPA is x

$$x + x + 0.25 = 6.71$$
$$-\ -0.25 = -0.25$$
$$\frac{2x}{2} = \frac{6.46}{2}$$
$$x = 3.23$$
$$3.23 + 0.25 = 3.48$$

The student's GPA is 3.48.

25.
$R \cdot T = D$
(60 mph) (3 hours) = D
$(60)(3) = 180$
His parents live about 180 miles away.

26.
3:00 P.M. – 11:00 P.M.
8 hours travel time
$R \cdot T = D$
(55mph)(8 hours) = D
$(55)8 = 440$
Sylvia traveled 440 miles.

136

27.
$$P = 4s$$
$$64 = 4s$$
$$\frac{64}{4} = \frac{4s}{4}$$
$$16 = s$$

Each side is 16 feet

28.
$$C = 2\pi r$$
$$C = 2(3.142)5$$
$$C = 31.42$$
Circumference is approximately 31.42 inches.

29.
$$A = \pi r^2$$
$$A = (3.12)4^2$$
$$A = 50.27 \text{ inches}^2$$

The area is approximately 50.27 inches^2

30.
$$A = \frac{1}{2}bh$$
$$A = \frac{1}{2} \cdot 4 \cdot 3$$
$$A = 6\text{ft}^2$$

Area is 6ft^2

9.4 Using Equations in One Variable to Solve Coin and Ticket Problems

1. 2 dimes and 6 quarters

value	# of coins	Total
10	x	10x
25	8 − x	25(8 − x)

$$10x + 25(8 - x) = 170$$
$$10x + 200 - 25x = 170$$
$$-15x + 200 = 170$$
$$-200 = -200$$
$$\frac{-15x}{-15} = \frac{-30}{-15}$$
$$x = 2$$

2. 1 adult ticket and 6 childrens tickets

	value	# of tickets	total
adult	$12	x	12x
child	$8	7 − x	8(7 − x)

$$12x + 8(7 - x) = 60$$
$$12x + 56 - 8x = 60$$
$$4x + 56 = 60$$
$$-56 = -56$$
$$\frac{4x}{4} = \frac{4}{4}$$
$$x = 1$$

3.   4 nickels and 8 dimes

value	# of coins	Total
5	x	5x
10	2x	10(2 x)

$$5x + 10(2x) = 100$$
$$5x + 20x = 100$$
$$\frac{25x}{25} = \frac{100}{25}$$
$$x = 4$$

4.   85 adults and 45 children

	value	# of tickets	total
adult	110	x	110x
child	55	130 - x	55(130 - x)

$$110x + 55(130 - x) = 11{,}825$$
$$110x + 7150 - 55x = 11{,}825$$
$$55x + 7150 = 11{,}825$$
$$-7150 = -7150$$
$$\frac{55x}{55} = \frac{4675}{55}$$
$$x = 85$$

5.   5 children are under 12; 3 children are 12 or older

	value	# of tickets	total
adult	195	x	195x
children	95	8 - x	95(8 - x)

$$195x + 95(8 - x) = 1060$$
$$195x + 760 - 95x = 1060$$
$$100x + 760 = 1060$$
$$-760 = -760$$
$$\frac{100x}{100} = \frac{300}{100}$$
$$x = 3$$

6.   11 3-cent stamps and 15 32-cent stamps

value	# of coins	Total
32	x + 4	32(x + 4)
3	x	3 x

$$32(x + 4) + 3x = 513$$
$$32x + 128 + 3x = 513$$
$$35x + 128 = 513$$
$$-128 = -128$$
$$\frac{35x}{35} = \frac{385}{35}$$
$$x = 11$$

7.   Answers will vary.

REVIEW

1.   Answers will vary.
2.   Answers will vary.
3.   Amswers will vary.

4.
$$R \cdot T = D$$
car $30(1 - x) = D$
train $60(x) = D$

$$60x + 30(1 - x) = 50$$
$$60x + 30 - 30x = 50$$
$$30x + 30 = 50$$
$$-30 = -30$$
$$\frac{30x}{30} = \frac{20}{30}$$
$$x = \frac{2}{3}$$

The train takes $\frac{2}{3}$ hour or 40 mins.

5.
$$R \cdot T = D$$
$$R \cdot 4\frac{36}{60} = 3247$$
$$R \cdot 4.6 = 3247$$
$$\frac{4.6R}{4.6} = \frac{3247}{4.6}$$

$R = 705.869$
Approximately 706 mph

6. Answers will vary.

7.
$$A = l \cdot w$$
$$A = 100 \cdot 60$$
$$A = 6000$$

$6,000\text{ft}^2$

8. Answers will vary.

9. California population 6x
Ohio population 2x
Indiana x

$$6x + 2x + x = 45,000,000$$
$$\frac{9x}{9} = \frac{45,000,000}{9}$$
$$x = 5,000,000$$

Indiana has 5,000,000
California has 10,000,000
California has 30,000,000

10. Answers will vary.

11. Answers will vary.

12. 73 adult tickets, 142 student tickets

	value	# of tickets	total
adult	3	x	3x
student	1	215 − x	1(215 − x)

$$3x + 1(215 - x) = 361$$
$$3x + 215 - x = 361$$
$$2x + 215 = 361$$
$$-215 = -215$$
$$\frac{2x}{2} = \frac{146}{2}$$
$$x = 73$$

13. Answers will vary.

PRACTICE TEST

1. Kent has 2x CDs
Jake has x CDs

$$2x + x = 150$$
$$\frac{3x}{3} = \frac{150}{3}$$
$$x = 50$$

Jake has 50 CDs, Kent has 100 CDs.

2.

Beginning $\frac{1}{2}$ tank,	$7\frac{1}{2}$ gallons
Put in	11 gallons
Total	$18\frac{1}{2}$ gallons
Left $\frac{1}{3}$ tank	5 gallons
Used	$13\frac{1}{2}$ gallons

Drove (21mpg)(13½ gallons)
(21)(13.5) = 283.5 miles
a. Used 13½ gallons
b. Drove 283.5 miles

3.
$$R \cdot T = D$$
$$60 \cdot T = 125 \text{ going}$$
$$\frac{60T}{60} = \frac{125}{60}$$
$$T = 2\frac{1}{12} \text{ hours} = 2 \text{ hrs } 5 \text{ mins}$$

$$R \cdot T = D$$
$$45 \cdot T = 125 \text{ returning}$$
$$\frac{45T}{45} = \frac{125}{45}$$
$$T = 2\frac{7}{9} \text{ hours} = 2 \text{ hrs } 47 \text{ mins}$$

Total time   4hrs 52 mins

4.

Boxes	4
4 x 3 boxes	12
12 x 2 boxes	24
Total	40

There are 40 boxes.

5.
$$A = l \cdot w$$
$$A = 19 \cdot 13$$
$$A = 247 \text{ft}^2$$

The Smiths will need 247ft² of carpet.

6. 5, 7, 4, 6, 3, 5, 2

5 + 2 = 7
7 − 3 = 4
4 + 2 = 6
6 − 3 = 3
3 + 2 = 5
5 − 3 = 2

7. B Y C X D W

A Z
B Y
C X
D W
E V
F U
B Y
C X
D W

8. "Guwo ro xutosbz es quwo ro koibp"
"Give me liberty or give me death"

9. If 35 exit the left door
100 - 35 = 65 exit the right door.

$$\frac{65}{100}(950) = 617.5$$

You will need at least 618 flyers.

140

10.

1st prize	$4x = \$200$
2nd prize	$2x = \$100$
3rd prize	$x = \$50$

$$x + 2x + 4x = 350$$
$$\frac{7x}{7} = \frac{350}{7}$$
$$x = 50$$

11. Answers will vary.

12. Ben has 6 quarters and 4 nickels

value	#coins	total
5	10 - x	5(10 - x)
25	x	25x

$$5(10 - x) + 25x = 170$$
$$50 - 5x + 25x = 170$$
$$50 + 20x = 170$$
$$-50 = -50$$
$$\frac{20x}{20} = \frac{120}{20}$$
$$x = 6$$

13.
$$P = 2l + 2w$$
$$600 = 2(2x) + 2(x)$$
$$600 = 4x + 2x$$
$$\frac{600}{6} = \frac{6x}{6}$$
$$100 = x$$

Width is 100 ft.

CUMULATIVE REVIEW: CHAPTERS 7, 8, AND 9

1. The symbol ≥ is used in mathematics for the phrase <u>Greater than or equal to</u>.

2. The symbol < is used to represent the phrase <u>Less than</u>.

3. If $6x + 9 = 7$, then the statement that $6x + 9 + (-9) = 7 + (-9)$ is an example of the <u>Addition Property of Equality</u>.

4. The number π is a/an <u>Irrational number</u>.

5. The <u>Distributive Property</u> states that, for all a, b, c ∈ Reals, a(b + c) = ab + ac and ab + ac = a(b + c).

6. If $7x = 14$, then the statement that

   $\frac{1}{7} \cdot 7x = 14 \cdot \frac{1}{7}$ is an example of the

   <u>Multiplication Property of Equality</u>.

7. If $-x < 6$, then the statement that $(-1)(-x) > (-1)(6)$ is an example of one of the <u>Multiplication Properties of Inequalities</u>.

8. Writing 0.00345 as 3.45×10^{-3} is an example of expressing a number in <u>Scientific notation</u>.

9. $0.23\overline{5}$ Is an example of a/an <u>Repeating decimal</u>.

10. The <u>Property of Reciprocals</u> states that,

    for all real numbers a(a ≠ 0), $a \cdot \frac{1}{a} = 1$.

11. Write 3.153×10^3 in standard form.
    315,300

12. Write 0.0000832 in scientific notation.
    8.32×10^{-5}

13. Express 0.015 as a fraction.

    $\frac{15}{1000} = \frac{3}{200}$

14. Express $0.\overline{3}$ as a fraction. $\frac{1}{3}$

15. Which is greater   0.6 or $\frac{2}{3}$

    0.6 or $0.\overline{6}$

    $0.\overline{6}$ or $\frac{2}{3}$ is greater

16. Express $\frac{2}{9}$ as a decimal. $2 \div 9 = 0.\overline{2}$

17. Express $\frac{3}{20}$ as a decimal.

    $3 \div 20 = 0.15$.

18. Express $5\frac{1}{4}$ as a decimal. 5.25

19. Round 0.461 to the nearest tenth.
    0.461 to the nearest tenth is 0.5.

20. Round 1.007 to the nearest hundredth.
    1.007 to the nearest hundredth is 1.01.

21. $0.402 \div 0.6 = 0.67$

22. $(2.95)(1.232) = 3.6344$

23. $\begin{array}{r} 159.73 \\ +\ \underline{5.71} \\ 165.44 \end{array}$

24. $27.83 - 165.894$

$$\begin{array}{r} -165.894 \\ +\ \underline{27.830} \\ -138.064 \end{array}$$

25.
$$0.9 + \frac{1}{3} - 2.5 = \frac{9}{10} + \frac{1}{2} - \frac{25}{10} =$$

$$\frac{27}{30} + \frac{10}{30} - \frac{75}{30} = \frac{27 + 10 + (-75)}{30} =$$

$$-\frac{38}{30} = -\frac{19}{15} = -1\frac{4}{15} = -1.2\overline{6}$$

26. $\dfrac{3}{5}(0.68 + 2.75) = 0.6(3.43) = 2.058$

27.
$$23 + a = \frac{36}{6}$$
$$23 + a = 6$$
$$\underline{-23 = -23}$$
$$a = -17$$

28.
$$x + 8 - 3 = 5 + x$$
$$x + 5 = 5 + x$$

Any real number

29.
$$-\frac{1}{5}y = 3$$
$$-\frac{5}{1}\left(-\frac{1}{5}y\right) = 3\left(-\frac{5}{1}\right)$$
$$y = -15$$

30.
$$7x = 56$$
$$\frac{7x}{7} = \frac{56}{7}$$
$$x = 8$$

31.
$$\begin{array}{rcl} 5x - 3 &=& 10 \\ 3 &=& 3 \\ \dfrac{5x}{5} &=& \dfrac{13}{5} \\ x &=& \dfrac{13}{5} \end{array}$$

32.
$$\begin{array}{rcl} 2n + 10 &=& 3n - 9 \\ -2n &=& -2n \\ 10 &=& n - 9 \\ 9 &=& 9 \\ 19 &=& n \end{array}$$

33.
$$\begin{array}{rcl} 2 - b &=& -8 - 3b \\ +3b &=& 3b \\ 2 + 2b &=& -8 \\ -2 &=& -2 \\ \dfrac{2b}{2} &=& \dfrac{-10}{2} \\ b &=& -5 \end{array}$$

34.
$$\begin{array}{rcl} -16 + 3a - 4 - 5a &=& 20 - 12a \\ -20 - 2a &=& 20 - 12a \\ 12a &=& 12a \\ -20 + 10a &=& 20 \\ 20 &=& 20 \\ \dfrac{10a}{10} &=& \dfrac{40}{10} \\ a &=& 4 \end{array}$$

35.
$$\begin{array}{rcl} -0.44y &=& 13.2 \\ \dfrac{-44y}{-44} &=& \dfrac{1320}{-44} \\ y &=& -30 \end{array}$$

143

36.
$$-2.3x - 2 = 1.7x$$
$$-23x - 20 = 17x$$
$$23x \qquad = 23x$$
$$\frac{-20}{40} = \frac{40x}{40}$$
$$-\frac{1}{2} = x$$

37.
$$\frac{y}{4} + 1 = \frac{3}{8}$$
$$8(\frac{y}{4}) + 8(1) = 8(\frac{3}{8})$$
$$2y + 8 = 3$$
$$-8 = -8$$
$$\frac{2y}{2} = \frac{-5}{2}$$
$$y = -\frac{5}{2}$$

38.
$$\frac{1}{5}x - \frac{2}{3} = 7$$
$$5(\frac{1}{5}x) - 15(\frac{2}{3}) = 15(7)$$
$$3x - 10 = 105$$
$$10 = 10$$
$$\frac{3x}{3} = \frac{115}{3}$$
$$x = \frac{115}{3} = 38.\overline{3}$$

39.
$$y + 17 \leq 20$$
$$-17 \qquad -17$$
$$y \leq 3$$

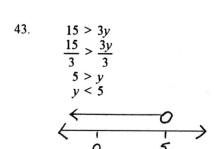

40.
$$-x + 5 > 3$$
$$-5 \quad -5$$
$$-x > -2$$
$$x < 2$$

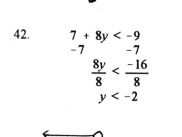

41.
$$2x \geq -12$$
$$\frac{2x}{2} \geq \frac{-12}{2}$$
$$x \geq -6$$

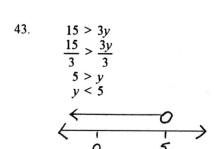

42.
$$7 + 8y < -9$$
$$-7 \qquad -7$$
$$\frac{8y}{8} < \frac{-16}{8}$$
$$y < -2$$

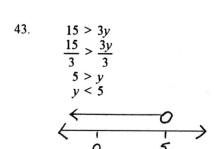

43.
$$15 > 3y$$
$$\frac{15}{3} > \frac{3y}{3}$$
$$5 > y$$
$$y < 5$$

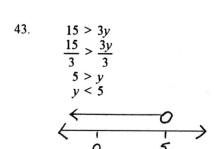

144

44.

$$R \cdot T = D$$
$$(65\text{mph})(3.5 \text{ hours}) = D$$
$$(65)(3.5) = 227.5$$

She has driven 227.5 miles.

45. 6600 ft. apart

Mary Ruth has run 3600 ft.

Meg has run 3000 ft.

$$R \cdot T = D$$
$$x(3 \text{ mins}) = 3000 \text{ ft}$$
$$\frac{x(3)}{3} = \frac{3000}{3}$$
$$x = 1000$$

She has a rate of 1000 feet per minute.

46.

49 hours, 100 points	= 2.5 GPA
25 hours, 72.5 points	= 2.9 GPA
100 points − 72.5 points	= 27.5 points
15 hours, 27.5 points	= $1.8\overline{3}$ GPA

He needs at least a $1.8\overline{3}$ GPA

47. A. 1, 3, 7, 15 <u>31</u>, <u>63</u>, <u>127</u>, 255

$$1 + 2 = 3$$
$$3 + 4 = 7$$
$$7 + 8 = 15$$
$$15 + 16 = 31$$
$$31 + 32 = 63$$
$$63 + 64 = 127$$
$$127 + 128 = 255$$

B.   6 dimes and 5 nickels

value	# coins	Total
5	11 − x	5(11 − x)
10	x	10x

$$10x + 5(11 - x) = 85$$
$$10x + 55 - 5x = 85$$
$$5x + 55 = 85$$
$$-55 = -55$$
$$\frac{5x}{5} = \frac{30}{5}$$
$$x = 6$$

145

CHAPTER 10: RATIO AND PROPOSITION

10.1 Ratios and Rates

1. 19 to 38 = $\dfrac{19}{38} = \dfrac{1}{2}$

2. 42 to 72 = $\dfrac{42}{72} = \dfrac{7}{12}$

3. 75 to 50 = $\dfrac{75}{50} = \dfrac{3}{2}$

4. 34 to 51 = $\dfrac{34}{51} = \dfrac{2}{3}$

5. 30 to 35 = $\dfrac{30}{35} = \dfrac{6}{7}$

6. 7 to 18 = $\dfrac{7}{18}$

7. 92 to 42 = $\dfrac{92}{42} = \dfrac{46}{21}$

8. 40 to 60 = $\dfrac{40}{60} = \dfrac{2}{3}$

9. 88 to 55 = $\dfrac{88}{55} = \dfrac{8}{5}$

10. 250 to 100 = $\dfrac{250}{100} = \dfrac{5}{2}$

11. 24 : 60 = $\dfrac{24}{60} = \dfrac{2}{5}$

12. 92 : 27 = $\dfrac{92}{27}$

13. 54 : 45 = $\dfrac{54}{45} = \dfrac{6}{5}$

14. 84 : 63 = $\dfrac{84}{63} = \dfrac{4}{3}$

15. 72 : 90 = $\dfrac{72}{90} = \dfrac{4}{5}$

16. $\dfrac{2}{3} : \dfrac{5}{6} + \dfrac{\frac{2}{3}}{\frac{5}{6}} =$

$$\dfrac{2}{3} \div \dfrac{5}{6} = \dfrac{2}{3} \cdot \dfrac{6}{5} = \dfrac{4}{5}$$

17. $\dfrac{3}{5}$ to $\dfrac{9}{10} = \dfrac{\frac{3}{5}}{\frac{9}{10}} =$

$$\dfrac{3}{5} \div \dfrac{9}{10} = \dfrac{3}{4} \cdot \dfrac{10}{9} = \dfrac{2}{3}$$

18. $2\dfrac{1}{2}$ to $1\dfrac{3}{4} = \dfrac{2\frac{1}{2}}{1\frac{3}{4}} =$

$$2\dfrac{1}{2} \div 1\dfrac{3}{4} = \dfrac{5}{2} \cdot \dfrac{4}{7} = \dfrac{10}{7}$$

19.

$$\frac{1}{2} \text{ to } 3\frac{1}{2} = \frac{\frac{1}{2}}{\frac{7}{2}} =$$

$$\frac{1}{2} \div \frac{7}{2} = \frac{1}{2} \cdot \frac{2}{7} = \frac{1}{7}$$

20.

$$\frac{5}{9} \text{ to } \frac{9}{10} = \frac{\frac{5}{9}}{\frac{9}{10}} =$$

$$\frac{5}{9} \div \frac{9}{10} = \frac{5}{9} \cdot \frac{10}{9} = \frac{50}{81}$$

21.

$$\frac{3}{4} \text{ to } \frac{3}{8} = \frac{\frac{3}{4}}{\frac{3}{8}} =$$

$$\frac{3}{4} \div \frac{3}{8} = \frac{3}{4} \cdot \frac{8}{3} = \frac{2}{1}$$

22.

$$4\frac{1}{8} \text{ to } 4\frac{1}{2} = \frac{\frac{33}{8}}{\frac{9}{2}} =$$

$$\frac{33}{8} \div \frac{9}{2} = \frac{33}{8} \cdot \frac{2}{9} = \frac{11}{12}$$

23.

$$\frac{7}{10} \text{ to } 3\frac{1}{2} = \frac{\frac{7}{10}}{\frac{7}{2}} =$$

$$\frac{7}{10} \div \frac{7}{2} = \frac{7}{10} \cdot \frac{2}{7} = \frac{1}{5}$$

24.

$$2\frac{1}{3} \text{ to } 1\frac{2}{3} = \frac{\frac{7}{3}}{\frac{5}{3}} =$$

$$\frac{7}{3} \div \frac{5}{3} = \frac{7}{3} \cdot \frac{3}{5} = \frac{7}{5}$$

25.

$$\frac{2}{7} \text{ to } \frac{4}{7} = \frac{\frac{2}{7}}{\frac{4}{7}} =$$

$$\frac{2}{7} \div \frac{4}{7} = \frac{2}{7} \cdot \frac{7}{4} = \frac{1}{2}$$

26.

$$\frac{2}{3} : \frac{7}{8} = \frac{\frac{2}{3}}{\frac{7}{8}} =$$

$$\frac{2}{3} \div \frac{7}{8} = \frac{2}{3} \cdot \frac{8}{7} = \frac{16}{21}$$

27.

$$\frac{3}{4} : \frac{1}{8} = \frac{\frac{3}{4}}{\frac{1}{8}} =$$

$$\frac{3}{4} \div \frac{1}{8} = \frac{3}{4} \cdot \frac{8}{1} = \frac{6}{1}$$

28.

$$3\frac{1}{2} : \frac{7}{8} = \frac{\frac{1}{32}}{\frac{7}{8}} =$$

$$\frac{7}{2} \div \frac{7}{8} = \frac{7}{2} \cdot \frac{8}{7} = \frac{4}{1}$$

29.

$$\frac{4}{5} : \frac{7}{10} = \frac{\frac{4}{5}}{\frac{7}{10}} =$$

$$\frac{4}{5} \div \frac{7}{10} = \frac{4}{5} \cdot \frac{10}{7} = \frac{8}{7}$$

30.

$$\frac{3}{19} : \frac{7}{38} = \frac{\frac{3}{19}}{\frac{7}{38}} =$$

$$\frac{3}{19} \div \frac{7}{38} = \frac{3}{19} \cdot \frac{38}{7} = \frac{6}{7}$$

31. $0.8 \text{ to } 6.4 = \frac{0.8}{6.4} = \frac{8}{64} = \frac{1}{8}$

32. $4 \text{ to } 3.2 = \frac{4}{3.2} = \frac{40}{32} = \frac{5}{4}$

33. $0.7 \text{ to } 21 = \frac{0.7}{21} = \frac{7}{210} = \frac{1}{30}$

34. $7 \text{ to } 2.1 = \frac{7}{2.1} = \frac{70}{21} = \frac{10}{3}$

35. $9.5 \text{ to } 3.8 = \frac{9.5}{3.8} = \frac{95}{38} = \frac{5}{2}$

36. $0.002 \text{ to } 2 = \frac{0.002}{2} = \frac{2}{2000} = \frac{1}{1000}$

37. $\frac{3}{4} \text{ to } 0.75 = 0.75 \text{ to } 0.75 = \frac{0.75}{0.75} = 1$

38. $1.25 \text{ to } \frac{1}{2} = \frac{1.25}{0.5} = \frac{125}{50} = \frac{5}{2}$

39. $3.7 \text{ to } \frac{7}{10} = \frac{3.7}{0.7} = \frac{37}{7}$

40. $0.6 \text{ to } 1.2 = \frac{0.6}{1.2} = \frac{6}{12} = \frac{1}{2}$

41. $4 : 0.04 = \frac{4}{0.04} = \frac{400}{4} = \frac{100}{1}$

42. $9.9 : 0.09 = \frac{9.9}{0.09} = \frac{990}{9} = \frac{110}{1}$

43. $0.75 : 2.5 = \frac{0.75}{2.5} = \frac{75}{250} = \frac{3}{10}$

44. $5.4 : 0.45 = \frac{5.4}{0.45} = \frac{540}{45} = \frac{12}{1}$

45. $0.49 : 7 = \frac{0.49}{7} = \frac{49}{700} = \frac{7}{100}$

46. $0.4 : \frac{1}{4} = \frac{0.4}{0.25} = \frac{40}{25} = \frac{8}{5}$

47. $\frac{7}{8} : 0.375 = \frac{0.875}{0.375} = \frac{875}{375} = \frac{7}{3}$

48. $0.5 : \frac{2}{3} = \frac{1}{2} \div \frac{2}{3} = \frac{1}{2} \cdot \frac{3}{2} = \frac{3}{4}$

49. $3.2 : \frac{7}{10} = \frac{3.2}{0.7} = \frac{32}{7}$

50. $\frac{3}{7} : \frac{5}{14} = \frac{3}{7} \div \frac{5}{14} = \frac{3}{7} \cdot \frac{14}{5} = \frac{6}{5}$

51. 2 feet to 2 inches

$$24 \text{ to } 2 = \frac{24}{2} = \frac{12}{1}$$

52. 15 minutes to 2 hours

$$15 \text{ to } 120 = \frac{15}{120} = \frac{1}{8}$$

148

53. 7 yards to 2 feet

21 to 2 $= \dfrac{21}{2}$

54. 4 hours to 45 minutes

240 to 45 $= \dfrac{240}{45} = \dfrac{16}{3}$

55. 10 seconds to 2 minutes

10 to 120 $= \dfrac{10}{120} = \dfrac{1}{12}$

56. 32 ounces to 2 pounds

32 to 32 $= \dfrac{32}{32} = 1$

57. 5,000 lbs to 4 tons

5,000 to 8,000 $= \dfrac{5000}{8000} = \dfrac{5}{8}$

58. 6 inches to 2 yards

6 to 72 $= \dfrac{6}{72} = \dfrac{1}{12}$

59. 5 minutes to 50 seconds

300 to 50 $= \dfrac{300}{50} \quad \dfrac{6}{1}$

60. 10 feet to 80 inches

120 to 80 $= \dfrac{120}{80} = \dfrac{3}{2}$

61. 3 weeks to 14 days

21 to 14 $= \dfrac{21}{14} = \dfrac{3}{2}$

62. 11 nickels to 2 dimes

11 to 4 $= \dfrac{11}{4}$

63. 6 pints to 5 quarts

6 to 10 $\dfrac{6}{10} = \dfrac{3}{5}$

64. 8 cents to 2 nickels

8 to 10 $= \dfrac{8}{10} \quad \dfrac{4}{5}$

65. 5 gallons to 5 quarts

20 to 5 $= \dfrac{20}{5} \quad \dfrac{4}{1}$

66. 10 ounces to 3 pints

10 to 48 $= \dfrac{10}{48} \quad \dfrac{5}{24}$

67. 3 pounds to 14 ounces

48 to 14 $= \dfrac{48}{14} \quad \dfrac{24}{7}$

68. 400 pounds to 2 tons

400 to 4,000 $= \dfrac{400}{4000} = \dfrac{1}{10}$

69. 2 hours to 19 minutes

120 to 19 $= \dfrac{120}{19}$

70. 14 inches to 3 feet

14 to 36 $= \dfrac{14}{36} \quad \dfrac{7}{18}$

71. 10 inches to 3 feet

$10 \text{ to } 36 = \dfrac{10}{36} = \dfrac{5}{18}$

72. 3 hours to 20 minutes

$180 \text{ to } 20 = \dfrac{180}{20} = \dfrac{9}{1}$

73. 4 feet to 3 yards

$4 \text{ to } 9 = \dfrac{4}{9}$

74. 50 minutes to 3 hours

$50 \text{ to } 180 = \dfrac{50}{180} = \dfrac{5}{18}$

75. 20 seconds to 3 minutes

$20 \text{ to } 180 = \dfrac{20}{180} \quad \dfrac{1}{9}$

76. 3 pounds to 14 ounces

$48 \text{ to } 14 = \dfrac{48}{14} = \dfrac{24}{7}$

77. 2 tons to 3,000 pounds

$4{,}000 \text{ to } 3{,}000 = \dfrac{4000}{3000} = \dfrac{4}{3}$

78. 9 inches to 3 yards

$9 \text{ to } 108 = \dfrac{9}{108} = \dfrac{1}{12}$

79. 45 seconds to 4 minutes

$45 \text{ to } 240 = \dfrac{45}{240} = \dfrac{3}{16}$

80. 7 inches to 4 feet

$7 \text{ to } 48 = \dfrac{7}{48}$

81. 5 days to 2 weeks

$5 \text{ to } 14 = \dfrac{5}{14}$

82. 3 quarters to 6 nickels

$15 \text{ to } 6 = \dfrac{15}{6} = \dfrac{5}{2}$

83. 7 dimes to 4 nickels

$14 \text{ to } 4 = \dfrac{14}{4} = \dfrac{7}{2}$

84. 3 pints to 3 quarts

$3 \text{ to } 6 = \dfrac{3}{6} = \dfrac{1}{2}$

85. 2 quarts to 3 gallons

$2 \text{ to } 12 = \dfrac{2}{12} = \dfrac{1}{6}$

86. 15 ounces to 2 pounds

$15 \text{ to } 48 = \dfrac{15}{48} = \dfrac{5}{16}$

87. 8 ounces to 2 pints

$8 \text{ to } 32 = \dfrac{8}{32} = \dfrac{1}{4}$

88. 250 pounds to 3 tons

$250 \text{ to } 6000 = \dfrac{250}{6000} = \dfrac{1}{24}$

89. 5 hours to 25 minutes

$$300 \text{ to } 25 = \frac{300}{25} = \frac{12}{1}$$

90. 7 inches to 5 feet

$$7 \text{ to } 60 = \frac{7}{60}$$

91. 5 month to 3 years

$$5 \text{ to } 36 = \frac{5}{36}$$

92. $$\frac{300 \text{ miles}}{5 \text{ hours}} = 60 \text{mph}$$

93. $$\frac{495 \text{ miles}}{9 \text{ hours}} = 55 \text{mph}$$

94. $$\frac{\$9086}{7 \text{ months}} = \$1298 \text{ per month}$$

95. $$\frac{\$173.70}{30 \text{ hours}} = \$5.79 \text{ per hour}$$

96. $$\frac{51 \text{ gallons}}{17 \text{ minutes}} = 3 \text{ gallons per minute}$$

97. $$\frac{299 \text{ students}}{13 \text{ sections}} = 23 \text{ students per section}$$

98. $$\frac{297 \text{ miles}}{11 \text{ gallons}} = 27 \text{mpg}$$

99. $$\frac{8 \text{ ounces}}{2 \text{ cubic feet}} = 4 \text{ ounces per foot}$$

100. Bottle $$\frac{\$2.56}{64 \text{ ounces}} = \$0.04 =$$

4 cents per ounce

Can $$\frac{\$1.44}{48 \text{ ounces}} = \$0.03 =$$

3 cents per ounce

The can is a better buy.

101. Jar 1. $$\frac{\$2.98}{28 \text{ ounces}} = \$0.016 \approx$$

10.6 cents per ounce

Jar 2. $$\frac{\$2.15}{20 \text{ ounces}} = \$0.107 \approx$$

10.7 cents per ounce

Jar 1 is a better buy.

102. $$\frac{260 \text{ miles}}{4 \text{ hours}} = 65 \text{mph}$$

103. $$\frac{2691 \text{ students}}{207 \text{ professors}} =$$

13 students per professor

104. $$\frac{\$141}{6 \text{ cu yds}} = \$23.50 \text{ per cubic foot}$$

105. $$\frac{518 \text{ miles}}{7 \text{ hours}} = 74 \text{mph}$$

106. $$\frac{708 \text{ miles}}{12 \text{ hours}} = 59 \text{mph}$$

151

107.
$$\frac{\$40,536}{12 \text{ months}} = \$3378 \text{ per month}$$

108.
$$\frac{\$142.56}{27 \text{ hours}} = \$5.28 \text{ per hour}$$

109.
$$\frac{65 \text{ gallons}}{26 \text{ minutes}} = 2.5 \text{ gpm}$$

110.
$$\frac{32 \text{ ounces}}{36 \text{ cu inches}} = 0.\overline{8} =$$

$\frac{8}{9}$ ounces per inch

111. Box 1
$$\frac{\$2.34}{18 \text{ ounces}} = \$0.13 =$$
13 cents per ounce

Box 2
$$\frac{\$2.21}{13 \text{ ounces}} = \$0.17 =$$
17 cents per ounce

Box 1 is a better buy.

112. Jar 1
$$\frac{\$3.68}{16 \text{ ounces}} = \$0.23 =$$
23 cents per ounce

Jar 2
$$\frac{\$3.38}{13 \text{ ounces}} = \$0.26 =$$
26 cents per ounce

Jar 1 is the better buy.

113.
$$\frac{33,912 \text{ students}}{1256 \text{ professors}} =$$
27 students per professor.

10.2 The Fundamental Property of Proposition

1.
$$\frac{1}{2} = \frac{5}{10}$$
$$10 = 10$$
True

2.
$$\frac{5}{8} = \frac{10}{16}$$
$$80 = 80$$
True

3.
$$\frac{27}{81} = \frac{33}{99}$$
$$2673 = 2673$$
True

4.
$$\frac{0.25}{\frac{3}{5}} = \frac{1}{2}$$
$$0.5 \neq 0.6$$
False

5.
$$\frac{0.04}{1.2} = \frac{1}{3}$$
$$0.12 \neq 1.2$$
False

6.

$$\frac{100}{\frac{1}{2}} = \frac{50}{\frac{1}{4}}$$

$$25 = 25$$
True

7.

$$\frac{x}{1} = \frac{12}{16}$$

$$6x = 12$$

$$\frac{6x}{6} = \frac{12}{6}$$

$$x = 2$$

8.

$$\frac{2}{5} = \frac{4}{x}$$

$$2x = 20$$

$$\frac{2x}{2} = \frac{20}{2}$$

$$x = 10$$

9.

$$\frac{3}{8} = \frac{9}{y}$$

$$3y = 72$$

$$\frac{3y}{3} = \frac{72}{3}$$

$$y = 24$$

10.

$$\frac{5}{a} = \frac{4}{8}$$

$$4a = 40$$

$$\frac{4a}{4} = \frac{40}{4}$$

$$a = 10$$

11.

$$\frac{18}{4} = \frac{y}{10}$$

$$4y = 180$$

$$\frac{4y}{4} = \frac{180}{4}$$

$$y = 45$$

12.

$$\frac{16}{12} = \frac{24}{n}$$

$$16n = 288$$

$$\frac{16n}{16} = \frac{288}{16}$$

$$n = 18$$

13.

$$\frac{30}{k} = \frac{12}{20}$$

$$12k = 600$$

$$\frac{12k}{12} = \frac{600}{12}$$

$$k = 50$$

14.

$$\frac{9}{z} = \frac{3}{11}$$

$$3z = 99$$

$$\frac{3z}{3} = \frac{99}{3}$$

$$z = 33$$

15.

$$\frac{5}{7} = \frac{5}{x}$$

$$5x = 35$$

$$\frac{5x}{5} = \frac{35}{5}$$

$$x = 7$$

16.

$$\frac{x}{10} = \frac{10}{4}$$

$$4x = 100$$

$$\frac{4x}{4} = \frac{100}{4}$$

$$x = 25$$

17.
$$\frac{a}{25} = \frac{18}{30}$$
$$30a = 450$$
$$\frac{30a}{30} = \frac{450}{30}$$
$$a = 15$$

18.
$$\frac{8}{9} = \frac{32}{m}$$
$$8m = 288$$
$$\frac{8m}{8} = \frac{288}{8}$$
$$m = 36$$

19.
$$\frac{35}{21} = \frac{x}{36}$$
$$21x = 1260$$
$$\frac{21x}{21} = \frac{1260}{21}$$
$$x = 60$$

20.
$$\frac{17}{51} = \frac{100}{c}$$
$$17c = 5100$$
$$\frac{17c}{17} = \frac{5100}{17}$$
$$c = 300$$

21.
$$\frac{x}{12} = \frac{10}{14}$$
$$14x = 120$$
$$\frac{14x}{14} = \frac{120}{14}$$
$$x = \frac{60}{7}$$

22.
$$\frac{a}{3} = \frac{15}{16}$$
$$16a = 45$$
$$\frac{16a}{16} = \frac{45}{16}$$
$$a = \frac{45}{16}$$

23.
$$\frac{40}{9} = \frac{y}{5}$$
$$9y = 200$$
$$\frac{9y}{9} = \frac{200}{9}$$
$$y = \frac{200}{9}$$

24.
$$\frac{25}{36} = \frac{x}{20}$$
$$36x = 500$$
$$\frac{36x}{36} = \frac{500}{36}$$
$$x = \frac{125}{9}$$

25.
$$\frac{18}{p} = \frac{4}{7}$$
$$4p = 126$$
$$\frac{4p}{4} = \frac{126}{4}$$
$$p = 31.5$$

26.
$$\frac{0.01}{0.1} = \frac{x}{100}$$
$$0.1x = 1$$
$$\frac{0.01x}{0.1} = \frac{1}{0.1}$$
$$x = 10$$

27.

$$\frac{\frac{1}{2}}{\frac{1}{4}} = \frac{y}{7}$$

$$\frac{1}{4}y = \frac{7}{2}$$

$$\frac{4}{1} \cdot \frac{1}{4}y = \frac{7}{2} \cdot \frac{4}{1}$$

$$y = 14$$

28.

$$\frac{\frac{2}{3}}{6} = \frac{x}{18}$$

$$6x = 12$$

$$\frac{6x}{6} = \frac{12}{6}$$

$$x = 2$$

29.

$$\frac{7}{23} = \frac{\frac{4}{5}}{x}$$

$$7x = \frac{4}{5} \cdot 23$$

$$\frac{1}{7} \cdot 7x = \frac{4}{5} \cdot 23 \cdot \frac{1}{7}$$

$$x = \frac{92}{35}$$

30.

$$\frac{5}{0.5} = \frac{50}{x}$$

$$5x = 25$$

$$\frac{5x}{5} = \frac{25}{5}$$

$$x = 5$$

31.

$$\frac{0.05}{1.2} = \frac{x}{6}$$

$$1.2x = 0.3$$

$$\frac{1.2x}{1.2} = \frac{0.3}{1.2}$$

$$x = 0.25$$

32.

$$\frac{x}{6} = \frac{3\frac{1}{2}}{5}$$

$$5x = 21$$

$$\frac{5x}{5} = \frac{21}{5}$$

$$x = \frac{21}{5}$$

33.

$$\frac{0.64}{0.4} = \frac{0.96}{y}$$

$$0.64y = 0.384$$

$$\frac{0.64y}{0.64} = \frac{0.384}{0.64}$$

$$y = 0.6$$

34.

$$\frac{0.5}{\frac{1}{4}} = \frac{n}{6}$$

$$\frac{1}{4}n = 3$$

$$\frac{4}{1} \cdot \frac{1}{4}n = 3 \cdot \frac{4}{1}$$

$$n = 12$$

35.

$$\frac{2\frac{1}{4}}{4} = \frac{x}{10}$$

$$4x = \frac{9}{4} \cdot 10$$

$$4x = \frac{45}{2}$$

$$\frac{1}{4} \cdot 4x = \frac{45}{2} \cdot \frac{1}{4}$$

$$x = \frac{45}{8}$$

36.

$$\frac{0.55}{0.72} = \frac{1.65}{m}$$

$$0.55m = 1.188$$

$$\frac{0.55m}{0.55} = \frac{1.188}{0.55}$$

$$m = 2.16$$

37.

$$\frac{0.6}{x} = \frac{1.2}{0.84}$$

$$1.2x = 0.504$$

$$\frac{1.2x}{1.2} = \frac{0.504}{1.2}$$

$$x = 0.42$$

38.

$$\frac{1.2}{2.7} = \frac{3.4}{x}$$

$$1.2x = 9.18$$

$$\frac{1.2x}{1.2} = \frac{9.18}{1.2}$$

$$x = 7.65$$

39.

$$\frac{1.8}{6.4} = \frac{x}{0.16}$$

$$64x = 0.288$$

$$\frac{6.4x}{6.4} = 0.288$$

$$x = 0.045$$

40.

$$\frac{x}{\frac{3}{4}} = \frac{10}{4}$$

$$4x = \frac{15}{2}$$

$$\frac{1}{4} \cdot 4x = \frac{15}{2} \cdot \frac{1}{4}$$

$$x = \frac{15}{8}$$

41.

$$\frac{9}{x} = \frac{7}{9}$$

$$7x = 81$$

$$\frac{7x}{7} = \frac{81}{7}$$

$$x = \frac{81}{7} = 11.57$$

42.

$$\frac{x}{13} = \frac{5}{9}$$

$$9x = 65$$

$$\frac{9x}{9} = \frac{65}{9}$$

$$x = 7.\overline{2}$$

43.

$$\frac{3}{11} = \frac{x}{7}$$

$$11x = 21$$

$$\frac{11x}{11} = \frac{21}{11}$$

$$x = \frac{21}{11}$$

44.

$$\frac{7}{9} = \frac{18}{x}$$

$$7x = 162$$

$$\frac{7x}{7} = \frac{162}{7}$$

$$x = \frac{162}{7}$$

45.

$$\frac{x}{5} = \frac{2\frac{1}{2}}{3}$$

$$3x = \frac{25}{3}$$

$$\frac{1}{3} \cdot 3x = \frac{25}{2} \cdot \frac{1}{3}$$

$$x = \frac{25}{6} = 4.1\overline{6}$$

46.

$$\frac{2.3}{6\frac{9}{10}} = \frac{6}{y}$$

$$2.3y = 41.4$$

$$\frac{2.3y}{2.3} = \frac{41.4}{2.3}$$

$$y = 18$$

47.

$$\frac{9\frac{1}{2}}{x} = \frac{2.6}{8}$$

$$2.6x = 76$$

$$\frac{2.6x}{2.6} = \frac{76}{2.6}$$

$$x = \frac{380}{13} = 29.23$$

48.

$$\frac{0.3}{y} = \frac{4.3}{12.9}$$

$$4.3y = 3.87$$

$$\frac{4.3y}{4.3} = \frac{3.87}{4.3}$$

$$y = 0.9$$

49.

$$\frac{x}{2\frac{1}{3}} = \frac{1.9}{\frac{4}{3}}$$

$$\frac{4}{5}x = \frac{7}{3} \cdot \frac{19}{10}$$

$$\frac{5}{4} \cdot \frac{4}{5}x = \frac{7}{3} \cdot \frac{19}{10} \cdot \frac{5}{4}$$

$$x = \frac{133}{24}$$

50.

$$\frac{\frac{1}{6}}{\frac{1}{3}} = \frac{x}{18}$$

$$\frac{1}{3}x = \frac{1}{6} \cdot 18$$

$$\frac{1}{3}x = 3$$

$$\frac{3}{1} \cdot \frac{1}{3}x = 3 \cdot \frac{3}{1}$$

$$x = 9$$

51.

$$\frac{5}{6} = \frac{15}{x}$$

$$5x = 90$$

$$\frac{5x}{5} = \frac{90}{5}$$

$$x = 18$$

52.

$$\frac{3}{4} = \frac{12}{x}$$

$$3x = 48$$

$$\frac{3x}{3} = \frac{48}{3}$$

$$x = 16$$

53.

$$\frac{2}{x} = \frac{6}{10}$$

$$6x = 20$$

$$\frac{6x}{6} = \frac{20}{6}$$

$$x = \frac{10}{3} = 3.\overline{3}$$

54.

$$\frac{a}{25} = \frac{2}{5}$$

$$5a = 50$$

$$\frac{5a}{5} = \frac{50}{5}$$

$$a = 10$$

55.

$$\frac{2}{7} = \frac{x}{7}$$

$$7x = 14$$

$$\frac{7x}{7} = \frac{14}{7}$$

$$x = 2$$

56.

$$\frac{y}{9} = \frac{24}{12}$$

$$12y = 216$$

$$\frac{12y}{12} = \frac{216}{12}$$

$$y = 18$$

57.

$$\frac{17}{51} = \frac{a}{12}$$

$$51a = 204$$

$$\frac{51a}{51} = \frac{204}{51}$$

$$a = 4$$

58.

$$\frac{15}{x} = \frac{7}{14}$$

$$7x = 210$$

$$\frac{7x}{7} = \frac{210}{7}$$

$$x = 30$$

59.

$$\frac{3}{12} = \frac{5}{x}$$

$$3x = 60$$

$$\frac{3x}{3} = \frac{60}{3}$$

$$x = 20$$

60.

$$\frac{15}{45} = \frac{7}{x}$$

$$15x = 315$$

$$\frac{15x}{15} = \frac{315}{15}$$

$$x = 21$$

10.3 Applications of Proposition

1.

$$\frac{\text{Miles } 252}{\text{Hours } 4} = \frac{x}{7}$$

$$4x = 1764$$

$$\frac{4x}{4} = \frac{1764}{4}$$

$$x = 441$$

441 miles

2.

$$\frac{\text{Hits } 22}{\text{Games } 16} = \frac{x}{56}$$

$$16x = 1232$$

$$\frac{16x}{16} = \frac{1232}{16}$$

$$x = 77$$

77 hits

158

3.
$$\frac{\text{Width } 3}{\text{Length } 4} = \frac{12}{x}$$
$$3x = 48$$
$$\frac{3x}{3} = \frac{48}{x}$$
$$x = 16$$

16 inches

4.
$$\frac{\text{Map } 2\frac{1}{2}}{\text{Miles } 1} = \frac{x}{5\frac{3}{4}}$$
$$x = \frac{5}{2} \cdot \frac{23}{4}$$
$$x = \frac{115}{8} = 14\frac{3}{8}$$

$14\frac{3}{8}$ inches on the map

5.
$$\frac{\text{Miles } 120}{\text{Minutes } 30} = \frac{x}{180}$$
$$30x = 21600$$
$$\frac{30x}{30} = \frac{21600}{30}$$
$$x = 720$$

720 miles

6.
$$\frac{\text{Water } 14}{\text{Antifreeze } 22} = \frac{x}{35}$$
$$22x = 490$$
$$\frac{22x}{22} = \frac{490}{22}$$
$$x = 22.\overline{27}$$

≈ 22.27 oz. of water

7.
$$\frac{\text{Feet } 2}{\text{Inches} \frac{1}{2}} = \frac{x}{6\frac{1}{2}}$$
$$\frac{1}{2}x = 13$$
$$\frac{2}{1} \cdot \frac{1}{2}x = 13 \cdot \frac{2}{1}$$
$$x = 26$$

Living room is 26 feet long.

$$\frac{\text{Feet } 2}{\text{Inch } \frac{1}{2}} = \frac{x}{2\frac{1}{2}}$$
$$\frac{1}{2}x = 5$$
$$\frac{2}{1} \cdot \frac{1}{2}x = 5 \cdot \frac{2}{1}$$
$$x = 10$$

Kitchen is 10 feet long.

$$\frac{\text{Feet } 2}{\text{Inch } \frac{1}{2}} = \frac{x}{6}$$
$$\frac{1}{2}x = 12$$
$$\frac{2}{1} \cdot \frac{1}{2}x = 12 \cdot \frac{2}{1}$$
$$x = 24$$

The garage is 24 feet wide.

$$\frac{\text{Feet } 2}{\text{Inch } \frac{1}{2}} = \frac{x}{3\frac{3}{4}}$$
$$\frac{1}{2}x = \frac{15}{2}$$
$$\frac{2}{1} \cdot \frac{1}{2}x = \frac{15}{2} \cdot \frac{2}{1}$$
$$x = 15$$

Hallway is 15 feet long.

8.
$$\frac{\text{Miles } 27}{\text{Gallons } 1} = \frac{2619}{x}$$
$$27x = 2619$$
$$\frac{27x}{27} = \frac{2619}{27}$$
$$x = 97$$
97 gallons

9.
$$\frac{\text{Yes votes } 340}{\text{Total votes } 1280} = \frac{x}{21760}$$
$$1280x = 7398400$$
$$\frac{1280x}{1280} = \frac{7398400}{1280}$$
$$x = 5780$$

5780 yes votes. The levy would fail.

10.
$$\frac{\text{Fertilizer lbs } 50}{\text{Square yards}} = \frac{x}{25,000}$$
$$5,000x = 25,000(50)$$
$$\frac{5000x}{5000} = \frac{25,000(50)}{5000}$$
$$x = 250$$
250 lb, five 50lb bags

11.
$$\frac{\text{Taxes } 875}{\text{Value of house } 132,500} = \frac{x}{53,000}$$
$$132500x = 875(53000)$$
$$\frac{132500x}{132500} = \frac{875(53000)}{132500}$$
$$x = 350$$
$350 taxes

12.
$$\frac{\text{Pages } 14}{\text{Minutes } 12} = \frac{203}{x}$$
$$14x = 2436$$
$$\frac{14x}{14} = \frac{2436}{14}$$
$$x = 174$$
174 minutes = 2.9 hours =
2 hrs. 54 minutes

13.
$$\frac{\text{Defective parts } 8}{\text{Total parts } 250} = \frac{x}{1875}$$
$$250x = 15000$$
$$\frac{250x}{250} = \frac{15000}{250}$$
$$x = 60$$
60 defective parts

14.
$$\frac{\text{Rolls of carpet}}{\text{Stairways } 3\frac{1}{2}} = \frac{x}{14}$$
$$3.5x = 14$$
$$\frac{3.5x}{3.5} = \frac{14}{3.5}$$
$$x = 4$$
4 rolls of carpet, none left over.

15.
$$\frac{\text{Gallons of gas } 1}{\text{Miles } 22} = \frac{x}{187}$$
$$22x = 187$$
$$\frac{22x}{22} = \frac{187}{22}$$
$$x = 8.5$$
$$12 - 8.5 = 3.5$$
3.5 gallons of gas left in the tank.

16.
$$\frac{\text{Tickets } 3}{\text{Cost } 23.25} = \frac{7}{x}$$
$$3x = 162.75$$
$$\frac{3x}{3} = \frac{162.75}{3}$$
$$x = 54.25$$

7 tickets cost $54.25.

160

17.
$$\frac{\text{Bricks } 102}{\text{Distance } 76\frac{1}{2}} = \frac{x}{191.25}$$
$$76.5x = 195{,}075$$
$$\frac{76.5x}{76.5} = \frac{195075}{76.5}$$
$$x = 255$$
255 bricks needed.
3 lots with 45 left over.

18.
$$\frac{\text{Ham } 3}{\text{People } 8} = \frac{x}{13}$$
$$8x = 39$$
$$\frac{8x}{8} = \frac{39}{8}$$
$$x = 4.875$$

0.875 of 16 oz. = 14 oz.
4 lb. 14 oz. Of ham to feed 13 people.

19.
$$\frac{\text{Skirts } 6}{\text{Fabric } 8} = \frac{21}{x}$$
$$6x = 168$$
$$\frac{6x}{6} = \frac{168}{6}$$
$$x = 28$$

28 yards of fabric

20.
$$\frac{\text{Gravel } 478}{\text{Length of path } 8} = \frac{x}{20}$$
$$8x = 9560$$
$$\frac{8x}{8} = \frac{9560}{8}$$
$$x = 1195$$

1195 lbs. of gravel

21.
$$\frac{\text{Wins } 26}{\text{Games } 44} = \frac{x}{154}$$
$$44x = 4004$$
$$\frac{44x}{44} = \frac{4004}{44}$$
$$x = 91$$

Win 91 games, lose 63 games

22.
$$\frac{\text{Yarn } \frac{3}{5}}{\text{Squares } 1} = \frac{x}{35}$$
$$x = 21$$

21 skeins of yarn

23. $450 \div 25 = 18$ lengths of the pool
$$\frac{\text{Lengths } 9}{\text{Minutes } 5} = \frac{18}{x}$$
$$9x = 90$$
$$\frac{9x}{9} = \frac{90}{9}$$
$$x = 10$$

10 minutes to swim 18 lengths

24.
$$\frac{\text{Hits } 7}{\text{Games } 12} = \frac{x}{60}$$
$$12x = 420$$
$$\frac{12x}{12} = \frac{420}{12}$$
$$x = 35$$

35 hits in 60 games

25.

$$\frac{\text{Miles } 7}{\text{Inches(map) } 3} = \frac{28}{x}$$
$$7x = 84$$
$$\frac{7x}{7} = \frac{84}{4}$$
$$x = 12$$

12 inches apart on the map.

26.

$$\frac{\text{Miles } 112}{\text{Hours } 4} = \frac{x}{\frac{1}{4}}$$
$$4x = 28$$
$$\frac{4x}{4} = \frac{28}{4}$$
$$x = 7$$

Travel 7 miles in 15 minutes

27.

$$\frac{\text{Chocolate chips } 2}{\text{Flour } 3} = \frac{1\frac{1}{2}}{x}$$
$$2x = 4.5$$
$$\frac{2x}{2} = \frac{4.5}{2}$$
$$x = 2.25$$

2.25 or 2¼ cups of flour

28.

$$\frac{\text{Miles } 285}{\text{Hours } 5} = \frac{x}{9}$$
$$5x = 2565$$
$$\frac{5x}{5} = \frac{2565}{5}$$
$$x = 513$$

513 miles in 9 hours.

29.

$$\frac{\text{Hits } 19}{\text{Games } 17} = \frac{x}{60}$$
$$17x = 1140$$
$$\frac{17x}{17} = \frac{1140}{17}$$
$$x = 67.06 = 67\frac{1}{17}$$

About 67 hits in 60 games

30.

$$\frac{\text{Width } 5}{\text{Length } 7} = \frac{x}{17.5}$$
$$7x = 87.5$$
$$\frac{7x}{7} = \frac{87.5}{7}$$
$$x = 12.5$$

The photo will be 12.5 inches wide.

31.

$$\frac{\text{Miles } 775}{\text{Minutes } 300} = \frac{x}{36}$$
$$300x = 27900$$
$$\frac{300x}{300} = \frac{27900}{300}$$
$$x = 93$$

Fly 93 miles in 36 minutes

32.

$$\frac{\text{Miles } 31}{\text{Gal. of gas } 1} = \frac{2635}{x}$$
$$31x = 2635$$
$$\frac{31x}{31} = \frac{2635}{31}$$
$$x = 85$$

85 gallons of gas

162

33.

$$\frac{\text{Yes votes } 546}{\text{Total votes } 798} = \frac{x}{75810}$$
$$798x = 41392260$$
$$\frac{798x}{798} = \frac{41392260}{798}$$
$$x = 51870$$

51,870 yes votes, the levy would pass.

34.

$$\frac{\text{Fertilizer } 50}{\text{Sq. yds. } 500} = \frac{x}{22,000}$$
$$500x = 1,100,000$$
$$\frac{500x}{500} = \frac{1100000}{500}$$
$$x = 2200$$

2,200 lbs. of fertilizer, 44 50-lb. bags

35.

$$\frac{\text{Taxes } 298}{\text{Value of house}} = \frac{298}{48,000} = \frac{x}{168,000}$$
$$48000x = 50064000$$
$$\frac{48000x}{48000} = \frac{50064000}{48000}$$
$$x = 1043$$

$1,043 in taxes

36.

$$\frac{\text{Pages } 15}{\text{Minutes } 13.5} = \frac{315}{x}$$
$$15x = 4252.5$$
$$\frac{15x}{15} = \frac{4252.5}{15}$$
$$x = 283.5$$

283½ minutes = 4.725 hours

37.

$$\frac{\text{Defective parts } 12}{\text{Total parts } 174} = \frac{x}{609}$$
$$174x = 7308$$
$$\frac{174x}{174} = \frac{7308}{174}$$
$$x = 42$$

42 defective parts

38.

$$\frac{\text{Tiles}}{\text{Classrooms}} \frac{7}{1\frac{1}{2}} = \frac{x}{15}$$
$$1.5x = 105$$
$$\frac{1.5x}{1.5} = \frac{105}{1.5}$$
$$x = 70$$

70 boxes, none left over.

39.

$$\frac{\text{Gals of gas } 1}{\text{Miles } 19} = \frac{x}{361}$$
$$19x = 361$$
$$\frac{19x}{19} = \frac{361}{19}$$
$$x = 19$$
$$21 - 19 = 2$$

2 gallons of gas left.

40.

$$\frac{\text{Tickets } 4}{\text{Cost } 35.16} = \frac{13}{x}$$
$$4x = 457.08$$
$$\frac{4x}{4} = \frac{457.08}{4}$$
$$x = 114.27$$

13 tickets cost $114.27.

41.

$$\frac{\text{Bricks } 79}{\text{Path length } 59\frac{1}{4}} = \frac{x}{231.75}$$

$$59.25x = 18308.25$$

$$\frac{59.25x}{59.25} = \frac{18308.25}{59.25}$$

$$x = 309$$

309 bricks, 4 lots with 91 left over.

42.

$$\frac{\text{Turkey } 5}{\text{People } 12} = \frac{x}{57}$$

$$12x = 285$$

$$\frac{12x}{12} = \frac{285}{12}$$

$$x = 23.75$$

0.75 of 12 oz. = 12 oz.

23 lbs 12 oz. of turkey for 57 people.

43.

$$\frac{\text{Skirts } 3}{\text{Fabric } 5} = \frac{36}{x}$$

$$3x = 180$$

$$\frac{3x}{3} = \frac{180}{3}$$

$$x = 60$$

60 yards of fabric

44.

$$\frac{\text{Gravel } 950}{\text{Path length } 12} = \frac{x}{30}$$

$$12x = 28500$$

$$\frac{12x}{12} = \frac{28500}{12}$$

$$x = 2375$$

2,375 lbs. of gravel

45.

$$\frac{\text{Wins } 42}{\text{Games } 50} = \frac{x}{225}$$

$$50x = 9450$$

$$\frac{50x}{50} = \frac{9450}{50}$$

$$x = 189$$

Win 189 games, lose 36 games

46.

$$\frac{\text{Skeins } \frac{3}{4}}{\text{Squares } 1} = \frac{x}{40}$$

$$x = \frac{3}{4} \cdot 40$$

$$x = 30$$

30 skeins

47. $875 \div 25 = 35$ lengths

$$\frac{\text{Lengths } 18}{\text{Minutes } 12} = \frac{35}{x}$$

$$18x = 420$$

$$\frac{18x}{18} = \frac{420}{18}$$

$$x = 23.\overline{3} = 23\frac{1}{3}$$

23⅓ minutes

48. It will take 5 minutes!

164

49.

$$\frac{\text{Small triangle } 4}{\text{Large triangle } 12} = \frac{3}{f}$$

$$4f = 36$$

$$\frac{4f}{4} = \frac{36}{4}$$

$$f = 9$$

$$\frac{\text{Small triangle } 4}{\text{Large triangle } 12} = \frac{5}{e}$$

$$4e = 60$$

$$\frac{4e}{4} = \frac{60}{4}$$

$$e = 15$$

50.

$$\frac{\text{Length } 15}{\text{Width } 9} = \frac{3}{x}$$

$$15x = 27$$

$$\frac{15x}{15} = \frac{27}{15}$$

$$x = 1.8 = 1\frac{4}{5} \text{ inches}$$

Width is 1.8 inches

REVIEW

1. Ratio, see Section 10.1
2. Rate, see Section 10.1
3. The Fundamental Property of Proportion, see Section 10.2

4. $17 \text{ to } 34 = \frac{17}{34} = \frac{1}{2}$

5. $75 \text{ to } 25 = \frac{75}{25} = \frac{3}{1}$

6. $\frac{3}{4} \text{ to } = \frac{\frac{3}{4}}{\frac{8}{3}} =$

$$\frac{3}{4} \div \frac{8}{3} = \frac{3}{3} \cdot \frac{3}{8} = \frac{9}{32}$$

7. $\frac{14}{5} \text{ to } \frac{7}{10} = \frac{\frac{14}{5}}{\frac{7}{10}} =$

$$\frac{14}{5} \div \frac{7}{10} = \frac{14}{5} \cdot \frac{10}{7} = \frac{4}{1}$$

8. $0.02 \text{ to } 2 = \frac{0.02}{2} = \frac{2}{200} = \frac{1}{100}$

9. $0.003 \text{ to } 30 = \frac{0.003}{30} =$

$$\frac{3}{30000} = \frac{1}{10000}$$

10. $0.9 : \frac{4}{5} = \frac{0.9}{0.8} = \frac{9}{8}$

11. $\frac{2}{3} : 0.4 = \frac{\frac{2}{3}}{\frac{4}{10}} =$

$$\frac{2}{3} \div \frac{4}{10} = \frac{2}{3} \cdot \frac{10}{4} = \frac{5}{3}$$

12. $10 \text{ inches to } 1 \text{ foot} = 10 : 12 =$

$$\frac{10}{12} = \frac{5}{6}$$

13. $2 \text{ yards to } 2 \text{ feet} = 6 : 2 =$

$$\frac{6}{2} = \frac{3}{1}$$

165

14. 16 ounces to 4 pounds = 16 : 64 =

$$\frac{16}{64} = \frac{1}{4}$$

15. 20 seconds to 2 minutes = 20 : 120 =

$$\frac{20}{120} = \frac{1}{6}$$

16. 3 hours to 30 minutes = 180 : 30 =

$$\frac{180}{30} \quad \frac{6}{1}$$

17. 5 pints to 2 gallons = 5 : 16 = $\frac{5}{16}$

18. 3 months to 7 years = 3 : 84 =

$$\frac{3}{84} = \frac{1}{28}$$

19. 2 quarts to 3 gallons = 2 : 12 =

$$\frac{2}{12} = \frac{1}{6}$$

20.
$$R \cdot T = D$$
$$R(11) = 550$$
$$11R = 550$$
$$\frac{11R}{11} = \frac{550}{11}$$
$$R = 50$$
50 mph.

21.
$$\frac{\$11,250}{9 \text{ months}} = 1250$$
$1,250 per month

22.
$$\frac{195\frac{1}{2} \text{ miles}}{8.5 \text{ gallons}} = 23$$
23 mpg.

23.
$$\frac{\$1.85}{5 \text{ lbs.}} = 0.37$$
37 cents per pound

24.
$$\frac{207 \text{ students}}{11 \text{ sections}} = 18.\overline{81}$$
19 students per section

25.
$$\frac{469.2 \text{ miles}}{17 \text{ gallons}} = 27.6$$
27.6 mpg.

26.
$$\frac{325 \text{ miles}}{5 \text{ hours}} = 65$$
65 mph.

27.
$$\frac{\$196}{8 \text{ cu. yds.}} = 24.5$$
$24.50 per cubic yard

28.
$$\frac{7}{14} = \frac{3}{x}$$
$$7x = 12$$
$$\frac{7x}{7} = \frac{42}{7}$$
$$x = 6$$

166

29.
$$\frac{17}{51} = \frac{20}{y}$$
$$17y = 1020$$
$$\frac{17y}{17} = \frac{1020}{17}$$
$$y = 60$$

30.
$$\frac{0.4}{1.2} = \frac{7}{x}$$
$$0.4x = 8.4$$
$$\frac{0.4x}{0.4} = \frac{8.4}{0.4}$$
$$x = 21$$

31.
$$\frac{\frac{3}{8}}{\frac{2}{3}} = \frac{a}{\frac{5}{3}}$$
$$\frac{2}{3}a = \frac{3}{8} \cdot \frac{5}{3}$$
$$\frac{3}{2} \cdot \frac{2}{3}a = \frac{3}{8} \cdot \frac{5}{3} \cdot \frac{3}{2}$$
$$a = \frac{15}{16}$$

32.
$$\frac{4}{9} = \frac{8}{x}$$
$$4x = 72$$
$$\frac{4x}{4} = \frac{72}{4}$$
$$x = 18$$

33.
$$\frac{5}{2} = \frac{y}{9}$$
$$2y = 45$$
$$\frac{2y}{2} = \frac{45}{2}$$
$$y = \frac{45}{2}$$

34.
$$\frac{35}{42} = \frac{7}{x}$$
$$35x = 294$$
$$\frac{35x}{35} = \frac{294}{35}$$
$$x = 8.4$$

35.
$$\frac{0.3}{0.4} = \frac{12}{x}$$
$$0.3x = 4.8$$
$$\frac{0.3x}{0.3} = \frac{4.8}{0.3}$$
$$x = 16$$

36.
$$\frac{\text{Apples } 3}{\text{Cost } 0.36} = \frac{11}{3}$$
$$3x = 3.96$$
$$\frac{3x}{3} = \frac{3.96}{3}$$
$$x = 1.32$$

11 lbs. Cost $1.32

37.
$$\frac{\text{Miles } 150}{\text{Minutes } 30} = \frac{x}{300}$$
$$30x = 45000$$
$$\frac{30x}{30} = \frac{45000}{30}$$
$$x = 1500$$

1500 miles

38.
$$\frac{\text{Length } 4}{\text{Width } 3} = \frac{10}{x}$$
$$4x = 30$$
$$\frac{4x}{4} = \frac{30}{4}$$
$$x = 7.5$$

7.5 inches wide

167

39.

$$\frac{\text{Miles } 432}{\text{Gas } 16} = \frac{513}{x}$$

$$432x = 8208$$

$$\frac{432x}{432} = \frac{8208}{432}$$

$$x \quad 19$$

19 gallons of gas

40.

$$\frac{\text{Map inches } 4}{\text{Miles } 1} = \frac{x}{13\frac{1}{2}}$$

$$x \quad 54$$

54 inches apart

41.

$$\frac{\text{Water } 20}{\text{Antifreeze } 28} = \frac{x}{70}$$

$$28x = 1400$$

$$\frac{28x}{28} = \frac{1400}{28}$$

$$x \quad 50$$

42.

$$\frac{\text{Fertilizer } 50}{\text{Sq. yds. } 5000} = \frac{x}{32000}$$

$$5000x = 1600000$$

$$\frac{5000x}{5000} = \frac{1600000}{5000}$$

$$x \quad 320$$

$320 \div 50 = 6.4$

320 lb., 7 bags

43.

$$\frac{\text{Taxes } 438}{\text{House value } 78,000} = \frac{x}{195,000}$$

$$78000x = 85410000$$

$$\frac{78000x}{78000} = \frac{95410000}{78000}$$

$$x \quad 1095$$

$1,095 in taxes

PRACTICE TEST

1. $42 : 72 = \dfrac{42}{72} = \dfrac{7}{12}$

2.

$$\frac{3}{4} \text{ to } \frac{7}{8} = \frac{\frac{3}{4}}{\frac{7}{8}} =$$

$$\frac{3}{4} \div \frac{7}{8} = \frac{3}{4} \cdot \frac{8}{7} = \frac{6}{7}$$

3. $40 \text{ to } 60 = \dfrac{40}{60} = \dfrac{2}{3} = 0.\overline{6}$

4.

$$4\frac{1}{8} \text{ to } 4\frac{1}{2} = \frac{\frac{33}{8}}{\frac{9}{2}} =$$

$$\frac{33}{8} \div \frac{9}{2} = \frac{33}{8} \cdot \frac{2}{9} = \frac{11}{12}$$

5. $\dfrac{3}{5} : \dfrac{9}{10} = \dfrac{0.6}{0.9} = \dfrac{6}{9} = \dfrac{2}{3} = 0.\overline{6}$

6.

$$2\frac{1}{3} : 1\frac{2}{3} = \dfrac{\frac{7}{3}}{\frac{5}{3}} =$$

$$\frac{7}{3} \cdot \frac{3}{5} = \frac{7}{5} = 1.4$$

7.   2 yards to 2 feet = 6 : 2 =

$$\frac{6}{2} = \frac{3}{1}$$

8.   2 minutes to 20 seconds = 20 : 20 =

$$\frac{120}{20} = \frac{6}{1} = 6$$

9.   5 cents to 2 dimes = 5 : 20 =

$$\frac{5}{20} = \frac{1}{4} = 0.25$$

10.   20 ounces to 2 pounds = 20 : 32 =

$$\frac{20}{32} = \frac{5}{8}$$

11.   2 gallons to 3 quarts = 8 : 3 =

$$\frac{8}{3} = 2.\overline{6}$$

12.   3 hours to 30 minutes = 180 : 30 =

$$\frac{180}{30} = \frac{6}{1}$$

13.

$$\frac{160 \text{ miles}}{4 \text{ hours}} = 40 \text{ mph.}$$

14.

15.

$$\frac{\$8584}{8 \text{ months}} = \$1073 \text{ per month}$$

$$\frac{164 \text{ miles}}{8 \text{ gals. of gas}} = 20.5 \text{ mpg.}$$

16.

$$\frac{\$2.40}{120 \text{ sheets}} = 0.02 = 2 \text{ cents per sheet}$$

$$\frac{\$2.25}{90 \text{ sheets}} = 0.025 =$$

2½ cents per sheet

The better buy is 120 sheets at $2.40

17.

$$\frac{2}{5} = \frac{4}{x}$$
$$2x = 20$$
$$\frac{2x}{2} = \frac{20}{2}$$
$$x \quad 10$$

18.

$$\frac{18}{4} = \frac{y}{10}$$
$$4y = 180$$
$$\frac{4y}{4} = \frac{180}{4}$$
$$y \quad 45$$

19.

$$\frac{\frac{1}{2}}{\frac{1}{4}} = \frac{x}{7}$$
$$\frac{1}{4}x = \frac{1}{2} \cdot \frac{7}{1}$$
$$\frac{4}{1} \cdot \frac{1}{4} = \frac{1}{2} \cdot \frac{7}{1} \cdot \frac{4}{1}$$
$$x = 14$$

20.
$$\frac{0.05}{1.2} = \frac{x}{6}$$
$$1.2x = 0.3$$
$$\frac{1.2x}{1.2} = \frac{0.3}{1.2}$$
$$x = 0.25$$

24.
$$\frac{\text{Lengths } 16}{\text{Minutes } 20} = \frac{x}{45}$$
$$20x = 720$$
$$\frac{20x}{20} = \frac{720}{20}$$
$$x = 36$$

36 lengths

21.
$$\frac{2\frac{1}{4}}{4} = \frac{x}{10}$$
$$4x = \frac{45}{2}$$
$$\frac{1}{4} \cdot 4x = \frac{45}{2} \cdot \frac{1}{4}$$
$$x = \frac{45}{8}$$

25.
$$\frac{\text{Pages } 15}{\text{Minutes } 28} = \frac{252}{x}$$
$$15x = 7056$$
$$\frac{15x}{15} = \frac{7056}{15}$$
$$x = 470.4$$

470.4 minutes

22.
$$\frac{\text{Hits } 25}{\text{Games } 18} = \frac{x}{54}$$
$$18x = 1350$$
$$\frac{18x}{18} = \frac{1350}{18}$$
$$x \quad 75$$

75 hits

23.
$$\frac{\text{Pansies } 3}{\text{Cost } 0.57} = \frac{11}{x}$$
$$3x = 6.27$$
$$\frac{3x}{3} = \frac{6.27}{3}$$
$$x = 2.09$$

11 pots would cost $2.09

170

CHAPTER 11: PERCENTS

11.1 Conversions: Fractions, Decimals and Percents

1. $\dfrac{2}{3} = \dfrac{2}{3} \cdot 100\% = 66\dfrac{2}{3}\%$

2. $\dfrac{7}{10} = \dfrac{7}{10} \cdot 100\% = 70\%$

3. $1\dfrac{1}{4} = \dfrac{5}{4} \cdot 100\% = 125\%$

4. $2\dfrac{1}{8} = \dfrac{17}{8} \cdot 100\% = 212.5\%$

5. $\dfrac{3}{8} = \dfrac{3}{8} \cdot 100\% = 37.5\%$

6. $4 = 4 \cdot 100\% = 400\%$

7. $1\dfrac{5}{6} = \dfrac{11}{6} \cdot 100\% = 183\dfrac{1}{3}\%$

8. $\dfrac{3}{16} = \dfrac{3}{16} \cdot 100\% = 18.75\%$

9. $0.37 = 0.37(100) = 37\%$

10. $0.62 = 0.62(100) = 62\%$

11. $0.14 = 0.14(100) = 14\%$

12. $3.1 = 3.1(100) = 310\%$

13. $0.07 = 0.07(100) = 7\%$

14. $0.002 = 0.002(100) = 0.2\%$

15. $1.06 = 1.06(100) = 106\%$

16. $4\% = 4 \div 100 = 0.04$

17. $3.5\% = 3.5 \div 100 = 0.035$

18. $5\dfrac{1}{4}\% = 5.25 \div 100 = 0.0525$

19. $11\dfrac{1}{2}\% = 11.5 \div 100 = 0.115$

20. $17\dfrac{3}{4}\% = 17.75 \div 100 = 0.1775$

21. $\dfrac{1}{2}\% = 0.5 \div 100 = 0.005$

22. $1\dfrac{1}{4}\% = 1.25 \div 100 = 0.0125$

23. $31\% = 31 \div 100 = 0.31$

24. $49\% = 49 \div 100 = 0.49$

25. $62\dfrac{1}{2}\% = 62.5 \div 100 = 0.625$

26. $112\% = 112 \div 100 = 1.12$

27. $124\frac{1}{2}\% = 124.5 \div 100 = 1.245$

28. $0.3\% = 0.3 \div 100 = 0.003$

29. $0.06\% = 0.06 \div 100 = 0.0006$

30. $8.1\% = 8.1 \div 100 = 0.081$

31. $8\% = \frac{8}{100} = \frac{2}{25}$

32. $6\% = \frac{6}{100} = \frac{3}{50}$

33. $15\% = \frac{15}{100} = \frac{3}{20}$

34. $24\% = \frac{24}{100} = \frac{6}{25}$

35. $30\% = \frac{30}{100} = \frac{3}{10}$

36. $42\% = \frac{42}{100} = \frac{21}{50}$

37. $112\% = \frac{112}{100} = 1\frac{12}{100} = 1\frac{3}{25}$

38. $37\frac{1}{2}\% = \frac{37.5}{100} = \frac{375}{1000} = \frac{3}{8}$

39. $7.5\% = \frac{7.5}{100} = \frac{75}{1000} = \frac{3}{40}$

40. $140\% = \frac{140}{100} = 1\frac{40}{100} = 1\frac{2}{5}$

41. $265\% = \frac{265}{100} = 2\frac{65}{100} = 2\frac{13}{20}$

42. $0.06\% = \frac{0.06}{100} = \frac{6}{10000} = \frac{3}{5000}$

43. $82\frac{1}{2}\% = \frac{82.5}{100} = \frac{825}{1000} = \frac{33}{40}$

44. $21\frac{1}{4}\% = \frac{21.25}{100} = \frac{2125}{10000} = \frac{17}{80}$

45. $6\frac{1}{4}\% = \frac{6.25}{100} = \frac{625}{10000} = \frac{1}{16}$

	Percent	Decimal	Fraction in lowest terms
46.	10%	0.1	$\frac{1}{10}$
47.	25%	0.25	$\frac{1}{4}$
48.	61%	0.61	$\frac{61}{100}$
49.	3%	0.03	$\frac{3}{100}$
50.	12%	0.12	$\frac{3}{25}$
51.	120%	1.2	$\frac{6}{5}$
52.	0.4%	0.004	$\frac{1}{250}$
53.	35%	0.35	$\frac{7}{20}$
54.	50%	0.5	$\frac{1}{2}$
55.	75%	0.75	$\frac{3}{4}$

172

	Percent	Decimal	Fraction in lowest terms
56.	0.5%	0.005	$\dfrac{1}{200}$
57.	80%	0.8	$\dfrac{4}{5}$
58.	80%	0.8	$\dfrac{4}{5}$
59.	6.5%	0.065	$\dfrac{13}{200}$
60.	125%	1.25	$\dfrac{5}{4}$
61.	5%	0.05	$\dfrac{1}{20}$
62.	$12\dfrac{1}{2}$%	0.125	$\dfrac{1}{8}$
63.	145%	1.45	$1\dfrac{9}{20}$
64.	7%	0.07	$\dfrac{7}{100}$
65.	25%	0.25	$\dfrac{1}{4}$
66.	20%	0.2	$\dfrac{1}{5}$
67.	67.5%	0.675	$\dfrac{27}{40}$
68.	$\dfrac{1}{2}$%	0.005	$\dfrac{1}{200}$
69.	$33\dfrac{1}{3}$%	$0.\overline{3}$	$\dfrac{1}{3}$

	Percent	Decimal	Fraction in lowest terms
70.	87.5%	0.875	$\dfrac{7}{8}$
71.	3.5%	0.035	$\dfrac{7}{200}$
72.	15%	0.15	$\dfrac{3}{20}$
73.	15%	0.15	$\dfrac{3}{20}$
74.	0.02%	0.0002	$\dfrac{1}{5000}$
75.	50%	0.5	$\dfrac{1}{2}$
76.	14%	0.14	$\dfrac{7}{50}$
77.	40%	0.4	$\dfrac{2}{5}$
78.	60%	0.6	$\dfrac{3}{5}$
79.	3.7%	0.037	$\dfrac{37}{1000}$
80.	1.2%	0.012	$\dfrac{3}{250}$
81.	0.12%	0.0012	$\dfrac{3}{2500}$

11.2 Basic Percent Problems or Cases of Percent

1. 50% of 90 is 45.

    Base: 90
    Rate: 50%
    Amount: 45

2. $33\frac{1}{3}\%$ of 60 is 20.

    Base: 60
    Rate: 33⅓%
    Amount: 20

3. 300% of 7 is 21.

    Base: 7
    Rate: 300%
    Amount: 21

4. 90% of 200 is 180.

    Base: 200
    Rate: 90%
    Amount: 180

5. 175% of 88 is 154.

    Base: 88
    Rate: 175%
    Amount: 154

6. 12 is 75% of 16.

    Base: 16
    Rate: 75%
    Amount: 12

7. 4.5 is 5% of 90.

    Base: 90
    Rate: 5%
    Amount: 4.5

8. 25 is 40% of 62.5.

    Base: 62.5
    Rate: 40%
    Amount: 25

9. 150% of 30 is 45.

    Base: 30
    Rate: 150%
    Amount: 45

10. 1/3 is 20% of 5/3.

    Base: 5/3
    Rate: 20%
    Amount: 1/3

11.
$$50\% \cdot 200 = x$$
$$0.5(200) = x$$
$$100 = x$$

12.
$$25\% \cdot 372 = x$$
$$0.25(372) = x$$
$$93 = x$$

13.
$$\frac{5}{100} = \frac{x}{20}$$
$$100x = 100$$
$$\frac{100x}{100} = \frac{100}{100}$$
$$x = 1$$

14.
$$\frac{x}{100} = \frac{19}{38}$$
$$38x = 1900$$
$$\frac{38x}{38} = \frac{1900}{38}$$
$$x = 50\%$$

15.
$$\frac{x}{100} = \frac{71}{284}$$
$$284x = 7100$$
$$\frac{284x}{284} = \frac{7100}{284}$$
$$x = 25\%$$

16.
$$\frac{x}{100} = \frac{92}{115}$$
$$115x = 9200$$
$$\frac{115x}{115} = \frac{9200}{115}$$
$$x = 80\%$$

17.
$$18 = 75\% \cdot x$$
$$18 = 0.75x$$
$$\frac{18}{0.75} = \frac{0.75x}{0.75}$$
$$24 = x$$

18.
$$21 = 60\% \cdot x$$
$$21 = 0.6x$$
$$\frac{21}{0.6} = \frac{0.6x}{0.6}$$
$$35 = x$$

19.
$$\frac{33\frac{1}{3}}{100} = \frac{23}{x}$$
$$\frac{100}{3}x = \frac{23}{x}$$
$$\frac{3}{100} \cdot \frac{100}{3}x = 2300 \cdot \frac{3}{100}$$
$$x = 69$$

20.
$$\frac{x}{100} = \frac{76}{400}$$
$$400x = 7600$$
$$\frac{400x}{400} = \frac{7600}{400}$$
$$x = 19\%$$

21.
$$\frac{x}{100} = \frac{4}{15}$$
$$15x = 400$$
$$\frac{15x}{15} = \frac{400}{15}$$
$$x = 26\frac{2}{3}\%$$

22.
$$\frac{x}{100} = \frac{7}{25}$$
$$25x = 700$$
$$\frac{25x}{25} = \frac{700}{25}$$
$$x = 28\%$$

23.
$$8\% \cdot 30 = x$$
$$0.08(30) = x$$
$$2.4 = x$$

24.
$$6 = x \cdot 27$$
$$6 = 27x$$
$$\frac{6}{27} = \frac{27x}{27}$$
$$0.\overline{2} = x$$
$$0.\overline{2} = 22.\overline{2}\%$$

25.
$$52 = x \cdot 39$$
$$52 = 39x$$
$$\frac{52}{39} = \frac{39x}{39}$$
$$1.\overline{3} = x$$
$$1.\overline{3} = 133.\overline{3}\% = 133\frac{1}{3}\%$$

26.
$$11\% \cdot 6 = x$$
$$0.11(6) = x$$
$$0.66 = x$$

27.
$$30\% \cdot 20 = x$$
$$0.3(20) = x$$
$$6 = x$$

28.
$$42\% \cdot 15 = x$$
$$0.42(15) = x$$
$$6.3 = x$$

29.
$$110\% \cdot 16 = x$$
$$1.1(16) = x$$
$$17.6 = x$$

30.
$$120\% \cdot 55 = x$$
$$1.2(55) = x$$
$$66 = x$$

175

31.

$$\frac{x}{100} = \frac{4}{\frac{1}{2}}$$

$$\frac{1}{2}x = 400$$

$$\frac{2}{1} \cdot \frac{1}{2}x = 400 \cdot 2$$

$$x = 800\%$$

32.

$$\frac{x}{100} = \frac{15}{12}$$

$$12x = 1500$$

$$\frac{12x}{12} = \frac{1500}{12}$$

$$x = 125\%$$

33.

$$120 = x \cdot 90$$

$$120 = 90x$$

$$\frac{120}{90} = \frac{90x}{90}$$

$$1.\overline{3} = x$$

$$1.\overline{3} = 133.\overline{3}\% = 133\frac{1}{3}\%$$

34.

$$\frac{x}{100} = \frac{140}{112}$$

$$112x = 14000$$

$$\frac{112x}{112} = \frac{14000}{112}$$

$$x = 125\%$$

35.

$$6\frac{1}{2}\% \cdot 36 = x$$

$$0.065(36) = x$$

$$2.34 = x$$

36.

$$7\frac{1}{2}\% \cdot 46 = x$$

$$0.075(46) = x$$

$$3.45 = x$$

37.

$$\frac{1}{4}\% \cdot 10 = x$$

$$0.25\%(10) = x$$

$$0.0025(10) = x$$

$$0.025 = x$$

38.

$$\frac{1}{5}\% \cdot 40 = x$$

$$0.2\%(40) = x$$

$$0.002(40) = x$$

$$0.08 = x$$

39.

$$40\% \cdot x = 6$$

$$0.4x = 6$$

$$\frac{0.4x}{0.4} = \frac{6}{0.4}$$

$$x = 15$$

40.

$$64\% \cdot x = 12$$

$$0.64x = 12$$

$$\frac{0.64x}{0.64} = \frac{12}{0.64}$$

$$x = 18.75$$

41.

$$\frac{33}{100} = \frac{55}{x}$$

$$33x = 5500$$

$$\frac{33x}{33} = \frac{5500}{33}$$

$$x = 166\frac{2}{3}$$

42.

$$30\% \cdot x = 60$$

$$0.3x = 60$$

$$\frac{0.3x}{0.3} = \frac{60}{0.3}$$

$$x = 200$$

43.
$$120\% \cdot x = 16$$
$$1.2x = 16$$
$$\frac{1.2x}{1.2} = \frac{16}{1.2}$$
$$x = 13.\overline{3} = 13\frac{1}{3}$$

44.
$$\frac{130}{100} = \frac{65}{x}$$
$$130x = 6500$$
$$\frac{130x}{130} = \frac{6500}{130}$$
$$x = 50$$

45.
$$\frac{x}{100} = \frac{\frac{1}{2}}{\frac{3}{4}}$$
$$\frac{3}{4}x = \frac{1}{2} \cdot 100$$
$$\frac{4}{3} \cdot \frac{3}{4}x = \frac{1}{2} \cdot 100 \cdot \frac{4}{3}$$
$$x = \frac{200}{3} = 66\frac{2}{3}\%$$

46.
$$\frac{x}{100} = \frac{\frac{7}{8}}{\frac{3}{4}}$$
$$\frac{3}{4}x = \frac{7}{8} \cdot 100$$
$$\frac{4}{3} \cdot \frac{3}{4}x = \frac{7}{8} \cdot 100 \cdot \frac{4}{3}$$
$$x = \frac{350}{3} = 116\frac{2}{3}\%$$

47.
$$0.2\% \cdot x = 10$$
$$0.002x = 10$$
$$\frac{0.002x}{0.002} = \frac{10}{0.002}$$
$$x = 5000$$

48.
$$0.4\% \cdot 25 = x$$
$$0.004(25) = x$$
$$0.1 = x$$

49.
$$5\frac{1}{4}\% \cdot x = 14$$
$$5.25\% \cdot x = 14$$
$$0.0525x = 14$$
$$\frac{0.0525x}{0.0525} = \frac{14}{0.0525}$$
$$x = 266\frac{2}{3}$$

50.
$$7\frac{1}{2}\% \cdot 40 = x$$
$$0.075(40) = x$$
$$3 = x$$

11.3 Percent Application Problems

1.
$$6\% \cdot \$45 = x$$
$$0.06(45) = x$$
$$2.7 = x$$
$$\$2.70 \text{ tax}$$

2.
$$\frac{\text{Defective parts}}{\text{Total}} \quad \frac{2}{100} = \frac{8}{x}$$
$$2x = 800$$
$$\frac{2x}{2} = \frac{800}{2}$$
$$x = 400$$
$$400 \text{ total parts}$$

3.
$$\frac{\text{Tax}\ 5}{\text{Price}\ 100} = \frac{15}{x}$$
$$5x = 1500$$
$$\frac{5x}{5} = \frac{1500}{3}$$
$$x = 300$$

The oven cost $300.

4.
$$\frac{\text{Commission}\ x}{\text{Cost}\ 100} = \frac{168}{800}$$
$$800x = 16800$$
$$\frac{800x}{800} = \frac{16800}{800}$$
$$x = 21\%$$

Commission rate is 21%

5.
$$\frac{\text{Raise}\ x}{\text{Salary}\ 100} = \frac{60}{1000}$$
$$1000x = 6000$$
$$\frac{1000x}{1000} = \frac{6000}{1000}$$
$$x = 6\%$$

6% raise

6.
$$\frac{\text{Tax}\ 4}{\text{Price}\ 100} = \frac{x}{350}$$
$$100x = 1400$$
$$\frac{100x}{100} = \frac{1400}{100}$$
$$x = 14$$

$14 tax

7.
$$\frac{\text{Tax}\ 6}{\text{Price}\ 100} = \frac{384}{x}$$
$$6x = 38400$$
$$\frac{6x}{6} = \frac{38400}{6}$$
$$x = 6400$$

Car cost $6,400

8.
$$\frac{\text{Tax}\ x}{\text{Price}\ 100} = \frac{27.50}{550}$$
$$550x = 2750$$
$$\frac{550x}{550} = \frac{2750}{550}$$
$$x = 5\%$$

5% tax rate

9.
$$\frac{\text{Commission}\ 7}{\text{Price}\ 100} = \frac{x}{85,000}$$
$$100x = 595000$$
$$\frac{100x}{100} = \frac{595000}{100}$$
$$x = 5950$$

$5,950 commission

10.
$$10\% \cdot x = \$32.50$$
$$0.1x = 32.5$$
$$\frac{0.1x}{0.1} = \frac{32.5}{0.1}$$
$$x = 325$$

Machine costs $325.

11.
$$\frac{\text{Commission}}{\text{Price}} \quad \frac{x}{100} = \frac{56}{700}$$
$$700x = 5600$$
$$\frac{700x}{700} = \frac{5600}{700}$$
$$x = 8\%$$

8% commission

12.
$$\frac{\text{Raise}}{\text{Salary}} \quad \frac{x}{100} = \frac{96}{1200}$$
$$1200x = 9600$$
$$\frac{1200x}{1200} = \frac{9600}{1200}$$
$$x = 8\%$$

8% raise

13.
$$\frac{\text{Increase}}{\text{Price}} \quad \frac{12.5}{100} = \frac{1562.50}{x}$$
$$12.5x = 156250$$
$$\frac{12.5x}{12.5} = \frac{156250}{12.5}$$
$$x = 12,500$$

Price of the car was $12,500.

14.
$$\frac{\text{Discount}}{\text{Price}} \quad \frac{15}{100} = \frac{x}{250}$$
$$100x = 3750$$
$$\frac{100x}{100} = \frac{3750}{100}$$
$$x = 37.50$$

Discount was $37.50.

15.
$$30\% \cdot x = \$36$$
$$0.3x = 36$$
$$\frac{0.3x}{0.3} = \frac{36}{0.3}$$
$$x = 120$$
Original price was $120.

16.
$$20\% \cdot \$39.50 = x$$
$$0.2(39.5) = x$$
$$7.9 = x$$
Tax is $7.90

17.
$$12\% \cdot \$28.45 = x$$
$$0.12(28.45) = x$$
$$3.414 = x$$
You save $3.41.

18.
$$15\% \cdot \$73,000 = x$$
$$0.15(73000) = x$$
$$10950 = x$$
You put down $10,950.

19.
$$\frac{\text{Profit}}{\text{Investment}} \quad \frac{x}{100} = \frac{720}{6000}$$
$$6000x = 72000$$
$$6000x6000 = \frac{72000}{6000}$$
$$x = 12\%$$

12% profit.

20.
$$8.67\% \cdot \$1,254 = x$$
$$0.0867(1254) = x$$
$$108.7218 = x$$
$108.72 will be deducted.

21. $12\% \cdot \$1200 = x$
    $0.12(1200) = x$
    $144 = x$
    Commission is $144.

22. $8\% \cdot \$17.50 = x$
    $0.08(17.5) = x$
    $1.4 = x$
    $1.40 tax

23. $5\% \cdot \$7500 = x$
    $0.05(7500) = x$
    $375 = x$
    Price is decreased by $375.

24. $\dfrac{\text{Hits } x}{\text{Pitches } 100} = \dfrac{48}{56}$
    $56x = 4800$
    $\dfrac{56x}{56} = \dfrac{4800}{56}$
    $x = 85.\overline{714285}$

    She hit 85.7%

25. $\dfrac{\text{Baskets } x}{\text{Tries } 100} = \dfrac{36}{48}$
    $48x = 3600$
    $\dfrac{48x}{48} = \dfrac{3600}{48}$
    $x = 75\%$

    He makes 75% of the baskets.

26. $28\% \cdot 60{,}000 = x$
    $0.28(60000) = x$
    $16800 = x$
    She will pay her lawyer $16,800.

27. $26\% \cdot \$20{,}000 = x$
    $0.26(20000) = x$
    $5200 = x$
    He will pay his lawyer $5,200.

28. $\dfrac{\text{Increase} \quad 11}{\text{Value of house} \quad 100} = \dfrac{x}{225{,}000}$
    $100x = 2475000$
    $\dfrac{100x}{100} = \dfrac{2475000}{100}$
    $x = 24{,}750$

    House value increases $24,750.

29. $\dfrac{\text{Increase} \quad 2\frac{1}{2}}{\text{House value} \quad 100} = \dfrac{x}{29{,}000}$
    $100x = 72500$
    $\dfrac{100x}{100} = \dfrac{72500}{100}$
    $x = 725$
    House value increases $725.

30. $x \cdot 24 \text{ hours} = 8 \text{ hours}$
    $24x = 8$
    $\dfrac{24x}{24} = \dfrac{8}{24}$
    $x = 0.\overline{3}$
    $0.\overline{3} = 33\frac{1}{3}\%$
    8 hours is 33⅓% of a day.

31. $12\frac{1}{2}\% \cdot 24 \text{ hours} = x$
    $0.125(24) = x$
    $3 = x$

    The movie is 3 hours long.

180

32.

$$\frac{\text{Part of hour}}{\text{Hour}} \quad \frac{x}{100} = \frac{12}{60}$$
$$60x = 1200$$
$$\frac{60x}{60} = \frac{1200}{60}$$
$$x = 20\%$$

20% of an hour.

33.

$$33\frac{1}{3}\% \cdot 60 \text{ minutes} = x$$
$$\frac{1}{3} \cdot 60 = 20$$

20 minutes is 33⅓% of an hour.

34.

$$\frac{\text{Months}}{\text{Year}} \quad \frac{x}{100} = \frac{3}{12}$$
$$12x = 300$$
$$\frac{12x}{12} = \frac{300}{12}$$
$$x = 25\%$$

3 months is 25% of a year.

35.

$$75\% \cdot 12 \text{ months} = x$$
$$0.75(12) = x$$
$$9 = x$$

9 months is 75% of a year.

36.

$$\frac{\text{Weeks}}{\text{Year}} \quad \frac{x}{100} = \frac{13}{52}$$
$$52x = 1300$$
$$\frac{52x}{52} = \frac{1300}{52}$$
$$x = 25\%$$

13 weeks is 25% of a year.

37.

$$75\% \cdot 52 \text{ weeks} = x$$
$$0.75(52) = x$$
$$39 = x$$

39 weeks are 75% of a year.

38.

$$\frac{\text{Voted}}{\text{Voters}} \quad \frac{x}{100} = \frac{6660}{18,000}$$
$$18000x = 666000$$
$$\frac{18000x}{18000} = \frac{666000}{18000}$$
$$x = 37\%$$

37% voted.

39.

$$6\% \cdot \$9850 = x$$
$$0.06(9850) = x$$
$$591 = x$$

$591 is overhead.

40.

$$\frac{\text{Absent}}{\text{Enrolled}} \quad \frac{x}{100} = \frac{7}{28}$$
$$28x = 700$$
$$\frac{28x}{28} = \frac{700}{28}$$
$$x = 25\%$$

25% were absent.

41.

$$\frac{\text{Absent}}{\text{Enrolled}} \quad \frac{20}{100} = \frac{9}{x}$$
$$20x = 900$$
$$\frac{20x}{20} = \frac{900}{20}$$
$$x = 45$$

45 students enrolled.

42.

$$\frac{\text{Tax } 6}{\text{Price } 100} = \frac{720}{x}$$

$$6x = 72000$$

$$\frac{6x}{6} = \frac{72000}{6}$$

$$x = 12,000$$

He paid $12,000 for his car.

43.

$$\frac{\text{Rent } x}{\text{Pay } 100} = \frac{1185}{3950}$$

$$3950x = 118500$$

$$\frac{3950x}{3950} = \frac{118500}{3950}$$

$$x = 30\%$$

Spends 30% on rent.

$$\frac{\text{Food } x}{\text{Pay } 100} = \frac{790}{3950}$$

$$3950x = 790000$$

$$\frac{3950x}{3950} = \frac{79000}{3950}$$

$$x = 20\%$$

Spends 20% on food.

$$\frac{\text{Clothing } x}{\text{Pay } 100} = \frac{592.50}{3950}$$

$$3950x = 59250$$

$$\frac{3950x}{3950} = \frac{59250}{3950}$$

$$x = 15\%$$

15% spent on clothing.

$$\frac{\text{Incidentals } x}{\text{Pay } \quad 100} = \frac{395}{3950}$$

$$3950x = 39500$$

$$\frac{3950x}{3950} = \frac{39500}{3950}$$

$$x = 10\%$$

10% spent on incidentals.

$$\frac{\text{Saved } x}{\text{Pay } 100} = \frac{711}{3950}$$

$$3950x = 71100$$

$$\frac{3950x}{3950} = \frac{71100}{3950}$$

$$x = 18\%$$

18% is saved.

$$\frac{\text{Entertainment } x}{\text{Pay} \quad 100} = \frac{276.50}{3950}$$

$$3950x = 27650$$

$$\frac{3950x}{3950} = \frac{27650}{3950}$$

$$x = 7\%$$

7% is spent on entertainment.

44.

$$\frac{\text{Tax } 10000}{\text{Cost } 100} = \frac{x}{0.10}$$

$$100x = 1000$$

$$\frac{100x}{100} = \frac{1000}{100}$$

$$x = 10$$

The tax on a 10 cent bullet would be $10.

11.4 Percent Increase and Decrease

1. Total cost is 100% + 6% tax = 106% of original price.

$$106\% \cdot \$45 = x$$

$$1.06(45) = x$$

$$47.7 = x$$

The sweater plus tax is $47.70.

182

2. $25 - 7 = 18$ were present.

$$\frac{x}{100} = \frac{18}{25}$$
$$25x = 1800$$
$$\frac{25x}{25} = \frac{1800}{25}$$
$$x = 72\%$$

72% were present.

3.   Total cost is 100% + 20% tax = 120%.

$$\frac{120}{100} = \frac{x}{39.50}$$
$$100x = 4740$$
$$\frac{100x}{100} = \frac{4740}{100}$$
$$x = 47.40$$

The pain plus tax is $47.40.

4.   Original price is 100% − 20% discount = 80%.

$$80\% \cdot \$60 = x$$
$$0.8(60) = x$$
$$48 = x$$
The new price is $48.

5.   Original price is 100% − 20% discount = 80%.
$$80\% \cdot \$19.95 = x$$
$$0.8(19.95) = x$$
$$15.96 = x$$
$15.96 is the new price.

The price is 100% + 6% tax = 106%
$$106\% \cdot \$15.96 = x$$
$$1.06(15.96) = x$$
$$16.9176 = x$$
The discount price and tax is $16.92

6. a.   Original price 100% − 30% discount = 70%
$$70\% \cdot \$18.50 = x$$
$$(0.7)(18.5) = x$$
$$12.95 = x$$
New price of the shirt is $12.95.

b. $$70\% \cdot \$124.75 = x$$
$$(0.7)(124.75) = x$$
$$87.325 = x$$
New price of the shoes is $87.33.

c. $$70\% \cdot \$14 = x$$
$$0.7(14) = x$$
$$9.8 = x$$
New price of the necktie is $9.80.

d. $$70\% \cdot \$2 = x$$
$$0.7(2) = x$$
$$1.4 = x$$
New price of the socks is $1.40.

7.   Actual increase is 2500 − 2000 = 500.

$$\frac{\text{Increase}}{\text{Original population}} \frac{x}{100} = \frac{500}{2000}$$
$$2000x = 50000$$
$$\frac{2000x}{2000} = \frac{50000}{2000}$$
$$x = 25\%$$

25% increase in population.

8.   Original cost 100% − 20% discount = 80%.
$$80\% \cdot \$45 = x$$
$$0.8(45) = x$$
$$36 = x$$
The new price is $36.00

9. Calculator plus tax is 100% + 8% = 108%.

$$108\% \cdot \$17.50 = x$$
$$1.08(17.50) = x$$
$$18.9 = x$$

The total price is $18.90.

10. Original price is 100% + 6% increase = 106%.

$$106\% \cdot \$9500 = x$$
$$1.06(9500) = x$$
$$10070 = x$$

New price of the car is $10,070.

11. Original price 100% + 4% increase = 104%.

$$104\% \cdot \$3.25 = x$$
$$1.04(3.25) = x$$
$$3.38 = x$$

The new price is $3.38 per pound.

12. New selling price is original price 100% + increase of 12.5% = 112.5%.

$$\frac{\text{Increase } 12.5}{\text{New price } 112.5} = \frac{1562.5}{x}$$
$$12.5x = 175781.25$$
$$\frac{12.5x}{12.5} = \frac{175781.25}{12.5}$$
$$x = 14062.5$$

The new price of the car is $14,062.50.

13. Actual decrease is $125 − $25.

$$\frac{\text{Decrease } \quad x}{\text{Original price } 100} = \frac{25}{125}$$
$$125x = 2500$$
$$\frac{125x}{125} = \frac{2500}{125}$$
$$x = 20$$

A 20% decrease in price.

14. Computer cost 100% + 6% tax = 106%.

$$\frac{\text{Total price } \quad 106}{\text{Original price } 100} = \frac{x}{3500}$$
$$100x = 371000$$
$$\frac{100x}{100} = \frac{371000}{100}$$
$$x = 3710$$

Total price of the computer is $3,710.

15. Actual increase is $25,000 − $20,000 = $5,000.

$$\frac{\text{Increase } \quad x}{\text{Original salary } 100} = \frac{5,000}{20,000}$$
$$20000x = 500000$$
$$\frac{20000x}{20000} = \frac{500000}{20000}$$
$$x = 25\%$$

Percent increase of 25%.

16. Original price is 100% − 15% discount = 85%.

$$85\% \cdot \$110 = x$$
$$0.85(110) = x$$
$$93.5 = x$$

She pays $93.50.

17. Shoes cost 100% + 5½% discount = 105.5%.

$$105.5\% \cdot \$40 = x$$
$$1.055(40) = x$$
$$42.2 = x$$

Shoes cost $42.20.

18. If 12% are freshmen, 88% are not freshmen.

$$88\% \cdot 2400 = x$$
$$0.88(2400) = x$$
$$2112 = x$$

2112 are not freshmen.

19. Actual discount is $80 − $64 = $16.

$$\frac{\text{Discount}}{\text{Original price}} \frac{x}{100} = \frac{16}{80}$$
$$80x = 1600$$
$$\frac{80x}{80} = \frac{1600}{80}$$
$$x = 20\%$$

Jodi received a 20% discount.

20. Her salary is 100% plus her raise of 8% = 108%.

$$108\% \cdot \$1200 = x$$
$$1.08(1200) = x$$
$$1296 = x$$

Her new salary is $1296 per month.

21. The caculator cost $12.98 − 0.98 = $12.00.

$$\frac{\text{Tax}}{\text{Cost}} \frac{x}{100} = \frac{0.98}{12.00}$$
$$12x = 98$$
$$\frac{12x}{12} = \frac{98}{12}$$
$$x = 8\frac{1}{6}\%$$

Tax rate is $8\frac{1}{6}\%$.

22.
$$\frac{\text{Chicken pox}}{\text{Enrollment}} \frac{15}{100} = \frac{90}{x}$$
$$15x = 9000$$
$$\frac{15x}{15} = \frac{9000}{15}$$
$$x = 600$$

600 students enrolled in the school.

23. Her salary is 100% + 8% increase = 108%.

$$108\% \cdot \$25,000 = x$$
$$1.08(25000) = x$$
$$27000 = x$$

Her new salary is $27,000

Her salary is 100% + 9% increase = 109%.

$$109\% \cdot \$102,000 = x$$
$$1.09(102000) = x$$
$$111,180 = x$$

Her new salary is $111,180.

185

25. $100\% - 22\% = 78\%$
$$\mathbf{78\%} \cdot \$450 = x$$
$$0.78(450) = x$$
$$351 = x$$
The new price is $351.

26. $100\% - 35\% = 65\%$
$$65\% \cdot \$99 = x$$
$$0.65(99) = x$$
$$64.35 = x$$
The new price is $64.35.

27. Actual discount is $24 - $18 = $6.

$$\frac{\text{Discount } x}{\text{Original cost } 100} = \frac{6}{24}$$
$$24x = 600$$
$$\frac{24x}{24} = \frac{600}{24}$$
$$x = 25\%$$

25% discount.

28. Actual decrease $125,000 - $115,000 = $10,000.

$$\frac{\text{Decrease } x}{\text{Original price } 100} = \frac{10,000}{125,000}$$
$$125,000x = 1000000$$
$$\frac{125000x}{125000} = \frac{1000000}{125000}$$
$$x = 8\%$$

An 8% decrease in price.

29. The bunnies cost the materials $100\% + 60\% = 160\%$.

$$\frac{\text{Materials, proft } 160}{\text{materials } 100} = \frac{15}{x}$$
$$160x = 1500$$
$$\frac{160x}{160} = \frac{1500}{160}$$
$$x = 9.375$$

The materials cost $9.38.

30. The tree costs $100\% + 40\%$ profit = 140%.

$$\frac{\text{Tree, profit } 140}{\text{Tree } 100} = \frac{126}{x}$$
$$140x = 12600$$
$$\frac{140x}{140} = \frac{12600}{140}$$
$$x = 90$$

The tree cost him $90.

31. Risen 550%, so now cost 650%
$$650\% \cdot (0.90) = x$$
$$6.5(0.9) = x$$
$$5.85 = x$$
Coffee now costs $5.85.

11.5 Simple Interest

1.
$$P \cdot R \cdot T = I$$
$$\$900(12\%)(1) = I$$
$$900(0.12)(1) = I$$
$$108 = I$$

$108 interest.

186

2.
$$P \cdot R \cdot T = I$$
$$\$14{,}000\,(7\%)\,(3) = I$$
$$14000\,(0.07)\,(3) = I$$
$$2940 = I$$

$2940 interest.

3.
$$P \cdot R \cdot T = I$$
$$P\,(5\%)\,(1) = \$4650$$
$$\frac{0.05P}{0.05} = \frac{4650}{0.05}$$
$$P = 93{,}000$$
$93,000 will be invested.

4.
$$P \cdot R \cdot T = I$$
$$P\,(7\%)\,(6\text{ months}) = \$280$$
$$\frac{0.035P}{0.035} = \frac{280}{0.035}$$
$$P = 8000$$
$8,000 will be invested.

5.
$$P \cdot R \cdot T = I$$
$$P\,(11\%)\,(3\text{ months}) = \$33$$
$$P\,(0.11)\,(0.25) = 33$$
$$\frac{0.0275P}{0.0275} = \frac{33}{0.0275}$$
$$P = 1200$$
$1200 will be invested.

6.
$$P \cdot R \cdot T = I$$
$$\$5400\,(R)\,(1) = \$648$$
$$\frac{5400R}{5400} = \frac{648}{5400}$$
$$R = 0.12 = 12\%$$
Invested at 12%

7.
$$P \cdot R \cdot T = I$$
$$\$650\,(R)\,(6\text{ months}) = \$19.50$$
$$650\,(R)\,(0.5) = 19.5$$
$$\frac{325R}{325} = \frac{19.5}{325}$$
$$R = 0.06 = 6\%$$

Invested at 6%.

8.
$$P \cdot R \cdot T = I$$
$$\$500\,(8\%)\,T = \$40$$
$$500\,(0.08)\,T = 40$$
$$\frac{40T}{40} = \frac{40}{40}$$
$$T = 1$$

He will need to be invested for 1 year.

9.
$$P \cdot R \cdot T = I$$
$$\$7500\,(9\%)\,T = \$168.75$$
$$7500\,(0.09)\,T = 168.75$$
$$\frac{675T}{675} = \frac{168.75}{675}$$
$$T = 0.25 = \frac{1}{4}$$

¼ of a year is 3 months.

10.
$$P \cdot R \cdot T = I$$
$$\$8000\,(5\tfrac{1}{2}\%)\,18\text{ months} = I$$
$$8000\,(0.055)\,(1.5) = I$$
$$660 = I$$

$660 interest.

11.

$$P \cdot R \cdot T = I$$
$$\$750\left(8\frac{1}{2}\%\right)(30 \text{ months}) = I$$
$$750(0.085)(2.5) = I$$
$$159.375 = I$$

$159.38 interest.

12.

$$P \cdot R \cdot T = I$$
$$\$1500(R)(3 \text{ months}) = \$15$$
$$1500(R)(0.25) = 15$$
$$\frac{375R}{375} = \frac{15}{375}$$
$$R = 0.04 = 4\%$$

4% interest rate.

13.

$$P \cdot R \cdot T = I$$
$$102,000(R)(2 \text{ months}) = \$935$$
$$102000 \cdot \frac{2}{12} \cdot R = 935$$
$$\frac{17000R}{17000} = \frac{935}{17000}$$
$$R = 0.055 = 5.5\% = 5\frac{1}{2}$$

5.5% or 5½% interest rate.

14.

$$P \cdot R \cdot T = I$$
$$79,000(R)(3 \text{ years}) = \$5,997.50$$
$$79000(3)(R) = 15997.50$$
$$\frac{237000R}{237000} = \frac{15997.50}{237000}$$
$$R = 0.0675 = 6.75\% = 6\frac{3}{4}$$

6¾% interest rate.

15.

$$P \cdot R \cdot T = I$$
$$P(8\%)(9 \text{ months}) = \$138$$
$$P(0.08)(0.75) = 138$$
$$\frac{0.06P}{0.06} = \frac{138}{0.06}$$
$$P = 2300$$

$2,300 was borrowed.

16.

$$P \cdot R \cdot T = I$$
$$\$2,300(4.5\%)(T) = \$51.75$$
$$2300(0.045)T = 51.75$$
$$\frac{103.5T}{103.5} = \frac{51.75}{103.5}$$
$$T = 0.5$$

0.5 of a year is 6 months.

17.

$$P \cdot R \cdot T = I$$
$$\$700(7\%)T = \$36.75$$
$$700(0.07)T = 36.75$$
$$\frac{49T}{49} = \frac{36.75}{49}$$
$$T = 0.75$$

0.75 of a year is 9 months.

18.

$$P \cdot R \cdot T = I$$
$$\$77,000\left(9\frac{1}{2}\%\right)T = \$3657.50$$
$$77000(0.095)T = 3657.50$$
$$\frac{7315T}{7315} = \frac{3657.50}{7315}$$
$$T = 0.5$$

0.5 of a year is 6 months.

188

19.
$$P \cdot R \cdot T = I$$
$$P(6\tfrac{1}{2}\%)(2 \text{ years}) = \$1235$$
$$P(0.065)(2) = 1235$$
$$\frac{0.13P}{0.13} = \frac{1235}{0.13}$$
$$P = 9500$$

$9,500 is invested.

20.
$$P \cdot R \cdot T = I$$
$$\$10,500(5\%)\,1 = I$$
$$10500(0.05)\,1 = I$$
$$525 = I$$

$525 interest. At the end of the year she will have $10,500 + $525 = $11,025.

21.
$$P \cdot R \cdot T = I$$
$$\$1,800(6\%)(3 \text{ months}) = I$$
$$1800(0.06)(0.25) = I$$
$$27 = I$$

$27 interest. At the end of 3 months he will have $1,800 + $27 = $1827.

22.
$$P \cdot R \cdot T = I$$
$$\$8,000(4\%)(9 \text{ months}) = I$$
$$8000(0.04)(0.75) = I$$
$$240 = I$$

$240 interst will be earned.

23.
$$P \cdot R \cdot T = I$$
$$\$7,500(6\%)(3 \text{ months}) = I$$
$$7500(0.06)(0.25) = I$$
$$112.5 = I$$

$112.50 interest will be earned.

24.
$$P \cdot R \cdot T = I$$
$$\$2300(5\tfrac{1}{2}\%)(3 \text{ years}) = I$$
$$2300(0.055)(3) = I$$
$$379.5 = I$$

$379.50 interest will be earned.

25.
$$P \cdot R \cdot T = I$$
$$\$3,300(9\%)(3 \text{ months}) = I$$
$$3300(0.09)(0.25) = I$$
$$74.25 = I$$

$74.25 interest will be owed on the loan.

26.
$$P \cdot R \cdot T = I$$
$$\$2,900(4.5\%)(15 \text{ months}) = I$$
$$2900(0.045)(1.25) = I$$
$$163.125 = I$$

$163.13 interest will be owed on the loan.

27.
$$P \cdot R \cdot T = I$$
$$\$11,750(5\tfrac{1}{4}\%)(2 \text{ years}) = I$$
$$117500(0.0525)(2) = I$$
$$1233.75 = I$$

$1,233.75 interest will be earned.

28.
$$P \cdot R \cdot T = I$$
$$\$390(12\%)(12 \text{ months}) = I$$
$$390(0.12)(1) = I$$
$$46.8 = I$$

$46.80 interest will be earned.

29.

$$P \cdot R \cdot T = I$$
$$P(6\%)(18 \text{ months}) = \$540$$
$$P(0.06)(1.5) = 540$$
$$\frac{0.09P}{0.09} = \frac{540}{0.09}$$
$$P = 6000$$

$6,000 will be invested.

30.

$$P \cdot R \cdot T = I$$
$$\$8,000\left(8\frac{1}{4}\%\right)(6 \text{ months}) = I$$
$$8000(0.0825)(0.5) = I$$
$$330 = I$$

She would have $8,000 + $330 interest
$8,330.

31.

$$P \cdot R \cdot T = I$$
$$\$27,000(7\%)(1) = I$$
$$27000(0.07)1 = I$$
$$1890 = I$$

There would be $27,000 + $1,890
interest $28,890.

32.

$$P \cdot R \cdot T = I$$
$$\$8,000(6\%)1 = I$$
$$8000(0.06)1 = I$$
$$480 = I$$

The student will owe $8,000 + $480 =
$8,480.

33.

$$P \cdot R \cdot T = I$$
$$\$500(6\%)1 = I$$
$$500(0.06)1 = I$$
$$30 = I$$

Stacie will have $500 + $30 = $530.

$$P \cdot R \cdot T = I$$
$$\$450(12\%)1 = I$$
$$450(0.12)1 = I$$
$$54 = I$$

Steve will have $450 + $54 = $504.
Stacie will have a larger balance.

REVIEW

1. Percent, see Section 11.1
2. Base, see Section 11.1
3. Rate, see Section 11.1
4. Amount, see Section 11.1
5. Simple interst, see Section 11.5
6. Principal, see Section 11.5

7. $40\% = \dfrac{40}{100} = \dfrac{2}{5}$

8. $4\% = \dfrac{4}{100} = \dfrac{1}{25}$

9. $33\dfrac{1}{3}\% = \dfrac{33\frac{1}{3}}{100} = \dfrac{\frac{100}{3}}{100} =$

$\dfrac{100}{3} \div 100 = \dfrac{100}{3} \cdot \dfrac{1}{100} = \dfrac{1}{3}$

10. $150\% = \dfrac{150}{100} = \dfrac{3}{2} = 1\dfrac{1}{2}$

11. $0.8\% = \dfrac{0.8}{100} = \dfrac{8}{1000} = \dfrac{1}{125}$

12. $\dfrac{1}{2} = \dfrac{1}{2} \cdot 100\% = 50\%$

13. $\dfrac{7}{8} = \dfrac{7}{8} \cdot 100\% = 87\dfrac{1}{2}\% = 87.5\%$

14. $\dfrac{2}{3} = \dfrac{2}{3} \cdot 100\% = 66\dfrac{2}{3}\%$

15. $\dfrac{3}{10} - \dfrac{3}{10} \cdot 100\% = 30\%$

16. $\dfrac{2}{9} = \dfrac{2}{9} \cdot 100\% = 22\dfrac{2}{9}\% = 22.\overline{2}\%$

17. $27\% = 27 \div 100 = 0.27$

18. $55\% = 55 \div 100 = 0.55$

19. $250\% = 250 \div 100 = 2.5$

20. $0.4\% = 0.4 \div 100 = 0.004$

21. $3\% = 3 \div 100 = 0.03$

22. $0.8 = 0.8(100\%) = 80\%$

23. $0.04 = 0.04(100\%) = 4\%$

24. $3.42 = 3.42(100\%) = 342\%$

25. $0.0346 = 0.0346(100\%) = 3.46\%$

26. $0.261 = 0.261(100\%) = 26.1\%$

27. $0.24 = 0.24(100\%) = 24\%$

28. $0.125 = 0.125(100\%) = 12.5\%$

29. $0.75 = 9.75(100\%) = 975\%$

30. $0.004 = 0.004(100\%) = 0.4\%$

31. 9 is 90% of what number?

$$9 = 0.9(x)$$
$$\dfrac{9}{0.9} = \dfrac{0.9x}{0.9}$$
$$10 = x$$

9 is 90% of 10.

32. What percent of 60 is 30?

$$\dfrac{x}{100} = \dfrac{30}{60}$$
$$\dfrac{60x}{60} = \dfrac{3000}{60}$$
$$x = 50$$
50% of 60 is 30.

33. 80% of what number is 28?

$$0.8(x) = 28$$
$$\dfrac{0.8x}{0.8} = \dfrac{28}{0.8}$$
$$x = 35$$

80% of 35 is 28.

34. What percent of 56 is 14?

$$\dfrac{x}{100} = \dfrac{14}{56}$$
$$\dfrac{56x}{56} = \dfrac{1400}{56}$$
$$x = 25$$
25% of 56 is 14.

35. 38% of 70 is what number?

$$0.38(70) = x$$
$$26.6 = x$$

38% of 70 is 26.6.

36. 0.5% of 200 is what number?

$$0.005(200) = x$$
$$1 = x$$

0.5% of 200 is 1.

37. 175% of 300 is what number?

$$1.75(300) = x$$
$$525 = x$$

175% of 300 is 525.

38. What percent is 16 of 48?

$$\frac{x}{100} = \frac{16}{48}$$
$$\frac{48x}{48} = \frac{1600}{48}$$
$$x = 33\frac{1}{3}$$

$33\frac{1}{3}$% is the percent of 16 of 48.

39. 37% of what number is 333?

$$0.37(x) = 333$$
$$\frac{0.37x}{0.37} = \frac{333}{0.37}$$
$$x = 900$$

37% of 900 is 333.

40. 0.25% of 3800 is what number?

$$\frac{0.25}{100} = \frac{x}{3800}$$
$$100x = 3800(0.25)$$
$$\frac{100x}{100} = \frac{950}{100}$$
$$x = 9.5$$

0.25% of 3800 is 9.5.

41.
$$6\frac{1}{2}\% \cdot \$300 = x$$
$$(0.065)(300) = x$$
$$19.5 = x$$

Tax of $19.50

42.
$$30\% \cdot \$275 = x$$
$$(0.3)(275) = x$$
$$82.5 = x$$

$82.50 would be saved.

43.
$$\frac{\text{Scored } x}{\text{Total } 100} = \frac{20}{25}$$
$$25x = 2000$$
$$\frac{25x}{25} = \frac{2000}{25}$$
$$x = 80\%$$

She scored 80%.

44.
$$\frac{\text{Discount } 25}{\text{Price } 100} = \frac{210}{x}$$
$$25x = 21000$$
$$\frac{25x}{25} = \frac{21000}{25}$$
$$x = 840$$

List price was $840.

45.

$$\frac{\text{Made } x}{\text{Tries } 100} = \frac{9}{15}$$
$$15x = 900$$
$$\frac{15x}{15} = \frac{900}{15}$$
$$x = 60\%$$

He made 60% of his tries.

46.

$$\frac{\text{Females } 52}{\text{Total } 100} = \frac{390}{x}$$
$$52x = 39000$$
$$\frac{52x}{52} = \frac{39000}{52}$$
$$x = 750$$

750 students taking mathematics.

47. Missed 5 of 25, has 20 correct.

$$\frac{\text{Correct } x}{\text{Total } 100} = \frac{20}{25}$$
$$25x = 2000$$
$$\frac{25x}{25} = \frac{2000}{25}$$
$$x = 80\%$$

She has 80% correct.

48. Actual discount is $65 - $52 = $13.

$$\frac{\text{Discount } x}{\text{Original price } 100} = \frac{13}{65}$$
$$65x = 1300$$
$$\frac{65x}{65} = \frac{1300}{65}$$
$$x = 20\%$$

A 20% discount

49. Cost is %100 - $25% discount = 75%.

$$\frac{\text{Pays } 75}{\text{Original cost } 100} = \frac{x}{84}$$
$$100x = 6300$$
$$\frac{100x}{100} = \frac{6300}{100}$$
$$x = 63$$

She pays $63 for the sweater.

50. The whole test is 100% - 80% = 20%.

$$\frac{\text{Missed } 20}{\text{Total } 100} = \frac{14}{x}$$
$$20x = 1400$$
$$\frac{20x}{20} = \frac{1400}{20}$$
$$x = 70$$

70 problems on the test.

51.

$$\frac{\text{Freshmen } 40}{\text{Total } 100} = \frac{18}{x}$$
$$40x = 1800$$
$$\frac{40x}{40} = \frac{1800}{40}$$
$$x = 45$$

45 students in the class.

52. Actual number absent 35 - 28 = 7.

$$\frac{\text{Absent } x}{\text{Total } 100} = \frac{7}{35}$$
$$35x = 700$$
$$\frac{35x}{35} = \frac{700}{35}$$
$$x = 20\%$$

20% of the students were absent.

53.
$$P \cdot R \cdot T = I$$
$$\$15{,}000\,(6\tfrac{1}{2}\%)\,(2\text{ years}) = I$$
$$15000\,(0.065)\,(2) = I$$
$$1950 = I$$
$1,950 interest.

54.
$$P \cdot R \cdot T = I$$
$$\$3{,}200\,(9\tfrac{1}{2}\%)\,(9\text{ months}) = I$$
$$3200\,(0.095)\,(0.75) = I$$
$$228 = I$$
$228 interest. He repayed $3,200 + $228 = $3,428.

55.
$$P \cdot R \cdot T = I$$
$$P\,(7\tfrac{1}{4}\%)\,(6\text{ months}) = \$1{,}305$$
$$P\,(0.0725)\,(0.5) = 1305$$
$$\frac{0.03625(P)}{0.03625} = \frac{1305}{0.03625}$$
$$P = 36{,}000$$
$36,000 was invested.

56.
$$P \cdot R \cdot T = I$$
$$\$103{,}000\,(6\tfrac{1}{2}\%)\,T = \$3{,}347.50$$
$$103000\,(0.065)\,T = 3347.5$$
$$\frac{6695T}{6695} = \frac{3347.5}{6695}$$
$$T = 0.5$$
0.5 of a year is 6 months.

57.
$$P \cdot R \cdot T = I$$
$$\$785\,(R)\,(4\text{months}) = \$15.70$$
$$785\,(R)\,(\tfrac{4}{12}) = 15.7$$
$$\frac{785}{3}R = \frac{157}{10}$$
$$\frac{3}{785} \cdot \frac{785}{3}R = \frac{157}{10} \cdot \frac{3}{785}$$
$$R = 0.06 = 6\%$$
6% will earn $15.70 interest in 4 months.

58.
$$P \cdot R \cdot T = I$$
$$P\,(11\%)\,(3\text{ months}) = \$33$$
$$P\,(0.11)\,(0.25) = 33$$
$$\frac{0.0275P}{0.0275} = \frac{33}{0.0275}$$
$$P = 1200$$

$1,200 is all he can borrow.

59.
$$P \cdot R \cdot T = I$$
$$\$1{,}400\,(9\%)\,(T) = \$189$$
$$1400\,(0.09)\,T = 189$$
$$\frac{126T}{126} = \frac{189}{126}$$
$$T = 1.5$$

1.5 years is 1 year 6 months.

60.
$$P \cdot R \cdot T = I$$
$$\$7{,}000\,(R)\,(9\text{ months}) = \$236.75$$
$$7000\,(R)\,(0.75) = 236.75$$
$$\frac{5250R}{5250} = \frac{236.75}{5250}$$
$$R = 0.0450952$$

The rate of interest is 4.5%

61.

$$P \cdot R \cdot T = I$$

$$\$125{,}000 \left(8\frac{1}{2}\%\right)(30 \text{ months}) = I$$

$$125000\,(0.085)(2.5) = I$$

$$26562.5 = I$$

The investment will make \$26,562.50 in 30 months. There will be \$125,000 + \$26,562.50 = \$151,562.50 in the account.

PRACTICE TEST

	Fraction	Decimal	Percent
1.	$\frac{1}{4}$	0.25	25%
2.	$\frac{5}{8}$	0.625	$62\frac{1}{2}\%$
3.	$\frac{4}{5}$	0.8	80%
4.	$\frac{1}{3}$	$0.\overline{3}$	$33\frac{1}{3}\%$
5.	$1\frac{3}{4}$	1.75	175%
6.	$\frac{1}{250}$	0.004	0.4%
7.	$\frac{1}{200}$	0.005	0.5%
8.	$\frac{5}{8}$	0.625	62.5%
9.	$\frac{1}{200}$	0.005	$\frac{1}{2}\%$

10.

$$150\% \cdot 62 = x$$
$$1.5(62) = x$$
$$93 = x$$

93 is 150% of 62.

11.

$$\frac{25}{100} = \frac{32}{x}$$
$$25x = 3200$$
$$\frac{25x}{25} = \frac{3200}{25}$$
$$x = 128$$

32 is 25% of 128.

12.

$$\frac{x}{100} = \frac{76}{400}$$
$$400x = 7600$$
$$\frac{400x}{400} = \frac{7600}{400}$$
$$x = 19\%$$

76 is 19% of 400.

13.

$$\frac{135}{100} = \frac{54}{x}$$
$$135x = 5400$$
$$\frac{135x}{135} = \frac{5400}{135}$$
$$x = 40$$

135% of 40 is 54.

14.

$$36.75\% \cdot 28 = x$$
$$(0.3675)(28) = x$$
$$10.29 = x$$

36.75% of 28 is 10.29.

15.
$$\frac{x}{100} = \frac{1}{1000}$$
$$1000x = 100$$
$$\frac{1000x}{1000} = \frac{100}{1000}$$
$$x = \frac{1}{10}\% = 0.1\%$$
1 is 0.1% of 1000.

16.
$$15\% \cdot \$1,200 = x$$
$$(0.15)(1200) = x$$
$$180 = x$$
He will need \$180 for a down payment.

17.
$$\frac{\text{Hits } x}{\text{Pitches } 100} = \frac{36}{90}$$
$$90x = 3600$$
$$\frac{90x}{90} = \frac{3600}{90}$$
$$x = 40\%$$

The team hit 40% of the pitches.

18.
$$\frac{\text{Correct } 70}{\text{Total } 100} = \frac{63}{x}$$
$$70x = 6300$$
$$\frac{70x}{70} = \frac{6300}{70}$$
$$x = 90$$

There were 90 problems on the test.

19. New price is 100% + 8% = 108%.
$$\frac{\text{New price } 108}{\text{Original price } 100} = \frac{216}{x}$$
$$108x = 21600$$
$$\frac{108x}{108} = \frac{21600}{108}$$
$$x = 200$$

The TV originally cost \$200.

20. The actual decrease is \$5.00 – \$4.75 = \$0.25.
$$\frac{\text{Decrease } x}{\text{Original price } 100} = \frac{0.25}{5.00}$$
$$5x = 25$$
$$\frac{5x}{5} = \frac{25}{5}$$
$$x = 5\%$$

A 5% decrease in pay.

21. New population is 100% + 9% = 109%.
$$109\% \cdot 23,900 = x$$
$$1.09(23900) = x$$
$$26051 = x$$
The new population is 26,051.

22.
$$P \cdot R \cdot T = I$$
$$\$68,000 \left(15\frac{1}{2}\%\right)(6 \text{ months}) = I$$
$$68000(0.155)(0.5) = I$$
$$527 = I$$
Interest earned is \$527. \$6,800 + \$527 = \$7327 will be in the account.

23.
$$P \cdot R \cdot T = I$$
$$P\left(8\frac{1}{2}\%\right)(3 \text{ months}) = \$110.50$$
$$P(0.085)(0.25) = 110.5$$
$$\frac{0.02125P}{0.02125} = \frac{110.5}{0.02125}$$
$$P = 5200$$

$5,200 was invested.

24.
$$P \cdot R \cdot T = I$$
$$\$7,500\,(R)\,(2 \text{ years}) = \$825$$
$$7500\,(R)\,(2) = 825$$
$$\frac{15000R}{15000} = \frac{825}{15000}$$
$$R = 0.055 = 5.5\% = 5\frac{1}{2}\%$$

He will be invested at 5½%.

25.
$$P \cdot R \cdot T = I$$
$$\$13,000\,(4\%)\,T = \$390$$
$$13000\,(0.04)\,T = 390$$
$$\frac{520T}{520} = \frac{390}{520}$$
$$T = 0.75$$

0.75 of a year is 9 months.

CHAPTER 12: GEOMETRY AND MEASUREMENT

12.1 Linear Measure and Applications

1.
$$\frac{x \text{ inches}}{12 \text{ inches}} = \frac{4 \text{ ft.}}{1 \text{ ft.}}$$
$$x = 12(4)$$
$$x = 48 \text{ inches}$$

2.
$$\frac{x \text{ inches}}{12 \text{ feet}} = \frac{9.5}{1 \text{ foot}}$$
$$x = 12(9.5)$$
$$x = 114 \text{ inches}$$

3.
$$\frac{50 \text{ inches}}{12 \text{ inches}} = \frac{x \text{ feet}}{1 \text{ foot}}$$
$$12x = 50$$
$$\frac{12x}{12} = \frac{50}{12}$$
$$x = 4\frac{1}{6} = 4.1\overline{6} \text{ feet}$$

4.
$$\frac{103 \text{ inches}}{12 \text{ inches}} = \frac{x \text{ feet}}{1 \text{ foot}}$$
$$12x = 103$$
$$\frac{12x}{12} = \frac{103}{12}$$
$$x = 8\frac{7}{8} = 8.58\overline{3} \text{ feet}$$

5.
$$\frac{x \text{ feet}}{3 \text{ feet}} = \frac{5 \text{ yards}}{1 \text{ yard}}$$
$$x = 5(3)$$
$$x = 15 \text{ feet}$$

6.
$$\frac{x \text{ feet}}{3 \text{ feet}} = \frac{2.5 \text{ yards}}{1 \text{ yard}}$$
$$x = (2.5)3$$
$$x = 7.5 = 7\frac{1}{2} \text{ feet}$$

7.
$$\frac{75 \text{ feet}}{3 \text{ feet}} = \frac{x \text{ yards}}{1 \text{ yard}}$$
$$3x = 75$$
$$\frac{3x}{3} = \frac{75}{3}$$
$$x = 25 \text{ yards}$$

8.
$$\frac{23 \text{ feet}}{3 \text{ feet}} = \frac{x \text{ yards}}{1 \text{ yard}}$$
$$3x = 23$$
$$\frac{3x}{3} = \frac{23}{3}$$
$$x = 7\frac{2}{3} = 7.\overline{6} \text{ yards}$$

9.
$$\frac{x \text{ feet}}{5280 \text{ feet}} = \frac{2.5 \text{ miles}}{1 \text{ mile}}$$
$$x = 2.5(5280)$$
$$x = 13200 \text{ feet}$$

$$\frac{13200 \text{ feet}}{3 \text{ feet}} = \frac{x \text{ yards}}{1 \text{ yard}}$$
$$3x = 13200$$
$$\frac{3x}{3} = \frac{13200}{3}$$
$$x = 4400 \text{ yards}$$

10.
$$\frac{x \text{ feet}}{3 \text{ feet}} = \frac{3 \text{ yards}}{1 \text{ yard}}$$
$$x = 9 \text{ feet}$$

$$\frac{x \text{ inches}}{12 \text{ inches}} = \frac{9 \text{ feet}}{1 \text{ foot}}$$
$$x = 9(12)$$
$$x = 108 \text{ inches}$$

11.
$$\frac{x \text{ feet}}{5280 \text{ feet}} = \frac{1.75 \text{ miles}}{1 \text{ mile}}$$
$$x = 1.75\,(5280)$$
$$x = 9240 \text{ feet}$$

12.
$$\frac{95 \text{ inches}}{12 \text{ inches}} = \frac{x \text{ feet}}{1 \text{ foot}}$$
$$12x = 95$$
$$\frac{12x}{12} = \frac{95}{12}$$
$$x = 7\frac{11}{12} \text{ feet}$$

$$\frac{7\frac{11}{12} \text{ feet}}{3 \text{ feet}} = \frac{x \text{ yards}}{1 \text{ yard}}$$
$$3x = \frac{95}{12}$$
$$\frac{1}{3} \cdot 3x = \frac{95}{12} \cdot \frac{1}{3}$$
$$x = 2\frac{23}{36} = 2.63\overline{8} \text{ yards}$$

13. 12 meters = 1,200 cm.
14. 382 centimeters = 3.82 m
15. 15 meters = 15,000 mm
16. 3 meters = 300 cm
17. 3 meters = 3,000 mm
18. 5 kilometers = 5,000 m
19. 7.5 kilometers = 7,500 m
20. 853 meters = 0.853 km
21. 2,000 meters = 2 km
22. 93 millimeters = 9.3cm
23. 83 centimeters = 830mm
24. 748 centimeters = 7.48 m

25.
$$\frac{2.5 \text{ yards}}{1 \text{ yard}} = \frac{x \text{ feet}}{3 \text{ feet}}$$
$$x = 2.5\,(3)$$
$$x = 7.5 \text{ feet}$$

$$\frac{x \text{ inches}}{12 \text{ inches}} = \frac{7.5 \text{ feet}}{1 \text{ foot}}$$
$$x = 7.5\,(12)$$
$$x = 90 \text{ inches}$$

The door is 90 inches high.

26. 256 centimeters = 2.56 m
The carpet is 2.56 m wide.

27.
$$\frac{4 \text{ feet}}{1 \text{ foot}} = \frac{x \text{ inches}}{12 \text{ inches}}$$
$$x = 48 \text{ inches}$$
4 feet 5 inches =
48 inches + 5 inches = 53 inches.

28. 1.27 meters = 127 cm
The desk is 127 cm wide.

29. 13 centimeters = 130 mm
The computer desk is 130 mm wide.

30.
$$\frac{9 \text{ inches}}{12 \text{ inches}} = \frac{x \text{ feet}}{1 \text{ foot}}$$
$$12x = 9$$
$$12x12 = \frac{9}{12}$$
$$x = \frac{3}{4} = 0.75 \text{ feet}$$

The book is 0.75 ft. wide.

31. The line AB measures 3 inches.
32. The line CD measures 4½ inches.

33. The line EF measures 2¼ inches.
34. The line GH measures 3¾ inches.
35. The line JK measures 8 centimeters.
36. The line LM measures 3.5 centimeters.
37. The line NO measures 5 centimeters.
38. The line PQ measures 9 centimeters.
39. The line RS measures 5.5 centimeters.
40. The line TU measures 10 centimeters.

41.
$$c^2 = a^2 + b^2$$
$$c^2 = 4^2 + 3^2$$
$$c^2 = 16 + 9$$
$$c^2 = 25$$
$$c = \sqrt{25}$$
$$c = 5$$

42.
$$c^2 = a^2 + b^2$$
$$c^2 = 5^2 + 6^2$$
$$c^2 = 25 + 36$$
$$c^2 = 61$$
$$c = \sqrt{61}$$
$$c \approx 7.81024$$
$$c \approx 7.81$$

43.
$$c^2 = a^2 + b^2$$
$$c^2 = 8^2 + 6^2$$
$$c^2 = 64 + 36$$
$$c^2 = 100$$
$$c = \sqrt{100}$$
$$c = 10$$

44.
$$c^2 = a^2 + b^2$$
$$20^2 = a^2 + 16^2$$
$$400 = a^2 + 256$$
$$-256 = \qquad -256$$
$$144 = a^2$$
$$\sqrt{144} = a$$
$$12 = a$$

45.
$$c^2 = a^2 + b^2$$
$$15^2 = a^2 + 12^2$$
$$225 = a^2 + 144$$
$$-144 = \qquad -144$$
$$81 = a^2$$
$$\sqrt{81} = a$$
$$9 = a$$

46.
$$c^2 = a^2 + b^2$$
$$c^2 = 3^2 + 2^2$$
$$c^2 = 9 + 4$$
$$c^2 = 13$$
$$c = \sqrt{13}$$
$$c \approx 3.61$$

47.
$$c^2 = a^2 + b^2$$
$$16^2 = a^2 + 11^2$$
$$256 = a^2 + 121$$
$$-121 = \qquad -121$$
$$135 = a^2$$
$$\sqrt{135} = a$$
$$11.62 \approx a$$

48.
$$c^2 = a^2 + b^2$$
$$30^2 = a^2 + 18^2$$
$$900 = a^2 + 324$$
$$-324 = \qquad -324$$
$$576 = a^2$$
$$24 = a$$

49.

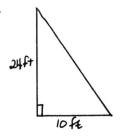

$c^2 = a^2 + b^2$
$c^2 = 24^2 + 10^2$
$c^2 = 576 + 100$
$c^2 = 676$
$c = 26$

The wire must be 26 feet long.

50.

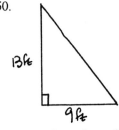

$c^2 = a^2 + b^2$
$c^2 = 13^2 + 9^2$
$c^2 = 169 + 81$
$c^2 = 250$
$c = \sqrt{250}$
$c \approx 15.81$

The ladder is approximately 15.81 m long.

51.

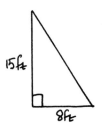

$c^2 = a^2 + b^2$
$c^2 = 15^2 + 8^2$
$c^2 = 225 + 64$
$c^2 = 289$
$c = \sqrt{289}$
$c = 17$

The ladder is 17 feet long.

52.

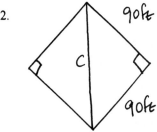

$c^2 = a^2 + b^2$
$c^2 = 90^2 + 90^2$
$c^2 = 8100 + 8100$
$c^2 = 16200$
$c = \sqrt{16200}$
$c \approx 127.28$

The distance is about 127.28 feet.

53.   p = a + b + c
      p = 9 + 10 + 8
      p = 27 in.

54.   p = a + b + c
      p = 10 + 10 + 5
      p = 25 cm

201

55. $p = 2l + 2w$
$p = 2(30) + 2(19)$
$p = 60 + 38$
$p = 98$ cm

56. $p = 2l + 2w$
$p = 2(15) + 2(9)$
$p = 30 + 18$
$p = 48$ in.

57. $p = 4s$
$p = 4(8)$
$p = 32$ ft.

58. 5 ft. = 60 in.
$p = 2l + 2w$
$p = 2(60) + 2(6)$
$p = 120 + 12$
$p = 132$ in.

59. $p = a + b + c + d + e$
$p = 8 + 8 + 6 + 6 + 6$
$p = 34$ m

60. 2 ft. = 24 ins.
$p = a + b + c + d$
$p = 24 + 16 + 20 + 14$
$p = 74$ in.

61. $p = 4s$
$p = 4(21)$
$p = 84$ ft.

62. $p = 4s$
$p = 4(2.3)$
$p = 9.2$ km

63. $p = 4s$
$p = 4(27)$
$p = 108$ mm

64. $p = 4s$
$p = 4(3\frac{1}{2})$
$p = 4 \cdot \frac{7}{2}$
$p = 14$ in.

65. $p = 2l + 2w$
$p = 2(8) + 2(7)$
$p = 16 + 14$
$p = 30$ ft.

66. $p = 2l + 2w$
$p = 2(90) + 2(3.5)$
$p = 180 + 7$
$p = 187$ mm

67. 3 ft. = 36 in.
$p = 2l + 2w$
$p = 2(4) + 2(36)$
$p = 8 + 72$
$p = 80$ in.

68. $p = 2l + 2w$
$p = 2(3) + 2(4)$
$p = 6 + 8$
$p = 14$ km

69. $c = \pi d$
$c \approx 3.14(3)$
$c \approx 9.42$ inches

70. $c = 2\pi 4$
$c \approx 2(3.14)8$
$c \approx 50.24$ feet

71. $c = \pi d$
$c \approx 3.14(2\frac{1}{2})$
$c \approx 3.14(2.5)$
$c \approx 7.85$ ft.

72. $c = 2\pi r$
$c \approx 2(3.14)(4.6)$
$c \approx 28.88$ mm

73.   c = πd
       c ≈ (3.14)7
       c ≈ 21.98cm

74.   c = 2πr
       c ≈ 2(3.14)5
       c ≈ 31.4m

75.   c = 2πr
       c ≈ 2(3.14)(6$\frac{1}{3}$)
       c ≈ 39.77 yd

76.   c = πd
       c ≈ (3.14)(8.5)
       c ≈ 26.69m

77.
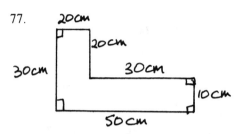

Perimeter is
20 + 20 + 30 + 10 + 50 + 30 = 160cm.

78.
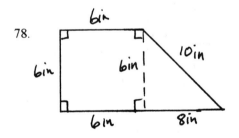

$c^2 = a^2 + b^2$
$c^2 = 8^2 + 6^2$
$c^2 = 64 + 36$
$c^2 = 100$
$c = \sqrt{100}$
$c = 10$

Perimeter is 10 + 6 + 6 + 14 = 36 inches.

79.
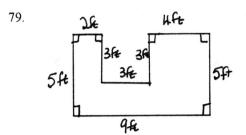

Perimeter is
2 + 3 + 3 + 3 + 4 + 5 + 9 + 5 = 34 feet.

80.
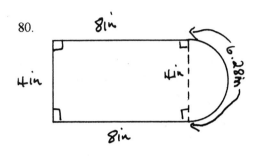

$c = πd$
$c ≈ (3.14)4$
$c ≈ 12.56$
$\frac{1}{2}c ≈ \frac{1}{2}(12.56)$
$\frac{1}{2}c ≈ 6.28$

Perimeter is
6.28 + 8 + 4 + 8 ≈ 26.28 inches.

203

81.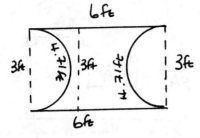

$c = \pi d$
$c \approx (3.14)3$
$c \approx 9.42$
$\frac{1}{2}c \approx (9.42)\frac{1}{2}$
$\frac{1}{2}c \approx 4.71$
Perimeter is
$4.71 + 6 + 4.71 + 6 \approx 21.42$ ft.

82.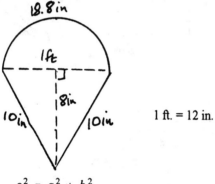

1 ft. = 12 in.

$c^2 = a^2 + b^2$
$c^2 = 8^2 + 6^2$
$c^2 = 64 + 36$
$c^2 = 100$
$c = \sqrt{100}$
$c = 10$

$c = \pi d$
$c \approx (3.14)12$
$c \approx 37.68$
$\frac{1}{2}c \approx (37.68)\frac{1}{2}$
$\frac{1}{2}c \approx 18.84$
Perimeter is
$10 + 10 + 18.84 \approx 38.84$ inches.

83.
$P = a + b + c$
$23 = 5 + 10 + c$
$23 = 15 + c$
$-15 = -15$
$8 = c$
$8\ cm.$

84.
$P = 4s$
$14 = 4s$
$\frac{14}{4} = \frac{4s}{4}$
$3.5 = s$
Each side has a length of 3.5 inches.

85.
$P = 2l + 2w$
$92 = 2(25) + 2w$
$-50 = -50$
$\frac{42}{2} = \frac{2w}{2}$
$21 = w$

The width is 21 feet.

86.
$c = \pi d$
$43.96 \approx 3.14d$
$\frac{43.96}{3.14} \approx \frac{3.14d}{3.14}$
$14 \approx d$
The diameter is ≈ 14 millimeters.

87.
$p = 2l + 2w$
$p = 2(21) + 2(17)$
$p = 42 + 34$
$p = 76$ inches
76 inches = $6.\overline{3}$ feet

$6\frac{1}{3}$ feet   or $6.\overline{3}$ feet   or

6 feet 4 inches of framing will be needed.

88.
$$p = 45$$
$$p = 4(12)$$
$$p = 48 \text{ feet } - 3 \text{ feet } = 45 \text{ feet}$$
45 feet of fencing will be needed.

89.
$$c = \pi d$$
$$c \approx (3.14)13$$
$$c \approx 40.82\text{m}$$
Approximately 40.82 meters of fencing will be needed.

90.
$$c = \pi d$$
$$25\frac{1}{2} \approx (3.14)d$$
$$25.5 \approx 3.14d$$
$$\frac{25.5}{3.14} \approx \frac{3.14d}{3.14}$$
$$8.12 \approx d$$
A circle of diameter approximately 8.12 feet.

91.
$$p = 2l + 2w$$
$$p = 2(9) + 2(4)$$
$$p = 18 + 8$$
$$p = 26$$

26 feet of Christmas decoration.

92.
$$c = \pi d$$
$$c \approx (3.14)4$$
$$c \approx 12.56 \text{ feet}$$
$$12.56 \text{ feet} \approx 4.19 \text{ yards}$$

≈ 4.19 yards of ribbon will be needed.

93.
$$p = 2l + 2w$$
$$p = 2(25) + 2(15)$$
$$p = 50 + 30$$
$$p = 80 \text{ yards}$$

$$\frac{\text{Feet } 3}{\text{Yard } 1} = \frac{x}{80}$$
$$x = 3(80)$$
$$x = 240 \text{ feet} \qquad p = 240 \text{ feet}$$

$$\frac{\text{Foot } 1}{\text{Inches } 12} = \frac{240}{x}$$
$$x = 240(12)$$
$$x = 2880 \text{ inches} \qquad p = 2880 \text{ inches}$$

1 tile each 6 inches
$2880 \div 6 = 480$ tiles.

94.
$$p = 2l + 2w$$
$$p = 2(10) + 2(9)$$
$$p = 20 + 18$$
$$p = 38$$

She will need 38 feet of border.

$$38 \div 15 = 2.5\overline{3}$$
She will need 3 rolls of border.

12.2 Square Measure and Applications

1.
$$\frac{288 \text{ inches}^2}{144 \text{ inches}^2} = \frac{x \text{ feet}^2}{1 \text{ foot}^2}$$
$$144x = 288$$
$$\frac{144x}{144} = \frac{288}{144}$$
$$x = 2 \text{ feet}^2$$

2.

$$\frac{1 \text{ yard}^s}{2 \text{ yards}^2} = \frac{9 \text{ feet}^2}{x \text{ feet}^2}$$
$$x = 9(2)$$
$$x = 18 \text{ feet}^s$$

3. $640 \text{ acres} = 1 \text{ mile}^2$

4.

$$\frac{1 \text{ yard}^2}{22 \text{ yards}^2} = \frac{9 \text{ feet}^2}{x \text{ feet}^2}$$
$$x = 9(22)$$
$$x = 198 \text{ feet}^2$$

5.

$$\frac{1 \text{ foot}^2}{40 \text{ feet}^s} = \frac{244 \text{ inches}}{x \text{ inches}}$$
$$x = 144(40)$$
$$x = 5760 \text{ inches}^2$$

6.

$$\frac{144 \text{ inches}^2}{36 \text{ inches}} = \frac{1 \text{ foot}^2}{x \text{ feet}^2}$$
$$144x = 36$$
$$\frac{144x}{144} = \frac{36}{144}$$
$$x = 0.25 \text{ feet}^2$$

7.

$$\frac{144 \text{ inches}^2}{1 \text{ inch}^2} = \frac{1 \text{ foot}^2}{x \text{ feet}^2}$$
$$144x = 1$$
$$\frac{144x}{144} = \frac{1}{144}$$
$$x = \frac{1}{144} \text{ feet}^2$$

8.

$$\frac{720 \text{ inches}^2}{144 \text{ inches}^2} = \frac{x \text{ feet}^2}{1 \text{ foot}^2}$$
$$144x = 720$$
$$\frac{144x}{144} = \frac{720}{144}$$
$$x = 5 \text{ feet}^2$$

$$\frac{5 \text{ feet}^2}{9 \text{ feet}^s} = \frac{x \text{ yards}^2}{1 \text{ yard}^2}$$
$$9x = 5$$
$$\frac{9x}{9} = \frac{5}{9}$$
$$x = \frac{5}{9} = 0.\overline{5} \text{ yards}^2$$

9.

$$\frac{3 \text{ yards}^2}{1 \text{ yard}^2} = \frac{x \text{ feet}^2}{9 \text{ feet}^2}$$
$$x = 3(9)$$
$$x = 27 \text{ feet}^2$$

$$\frac{27 \text{ feet}^2}{1 \text{ foot}^2} = \frac{x \text{ inches}^2}{144 \text{ inches}^2}$$
$$x = 144(27)$$
$$x = 3888 \text{ inches}^2$$

10.

$$\frac{2 \text{ acres}}{1 \text{ acres}} = \frac{x \text{ feet}^2}{43,560 \text{ feet}^2}$$
$$x = 2(43,560)$$
$$x = 87,120 \text{ feet}^2$$

11. $23 \text{cm}^2 = 2300 \text{mm}^2$

12. $200 \text{mm}^2 = 2 \text{cm}^2$

13. $10 \text{m}^2 = 100,000 \text{mm}^2$

14. $1.5 \text{m}^2 = 15,000 \text{cm}^2$

15. $1000 \text{mm}^2 = 0.001 \text{m}^2$

16. $5km^2 = 5,000,000m^2$

17. $9m^2 = 900dm^2$

18. $898mm^2 = 0.000898m^2$

19. $8m^2 = 80,000cm^2$

20. $A = l \cdot w$
$A = 40 \cdot 11$
$A = 440 \text{ inches}^2$

21. $A = 5^2$
$A = 15^2$
$A = 225 \text{ yards}^2$

22. $A = \frac{1}{2} \cdot b \cdot h$
$A = \frac{1}{2} \cdot \frac{13}{1} \cdot \frac{9}{1}$
$A = \frac{117}{2} = 58.5cm^2$

23. $A = \frac{1}{2} \cdot b \cdot h$
$A = \frac{1}{2} \cdot \frac{12}{1} \cdot \frac{8}{1}$
$A = 48cm^2$

24.
$4 \text{ feet} = 48 \text{ inches}$
$A = l \cdot w$
$A = 48(9)$
$A = 432 \text{ inches}^2$

25. $A = s^2$
$A = (3.4)^2$
$A = 11.56m^2$

26. $A = b \cdot h$
$A = 5 \cdot 3$
$A = 15cm^2$

27.
$A = \pi r^2$
$A \approx 3.14(3^2)$
$A \approx 3.14(9)$
$A \approx 28.26km^2$

28. $D = 8; 4 = 4$
$A = \pi r^2$
$A \approx 3.4(4^2)$
$A \approx 3.14(16)$
$A \approx 50.24 \text{ inches}^2$

29. $A = b \cdot h$
$A = 19(6)$
$A = 114 \text{ feet}^2$

30. $A = l \cdot w$
$A = 35(28)$
$A = 980 \text{ feet}^2$

$A = \frac{1}{2} \cdot b \cdot h$
$A = \frac{1}{2} \cdot \frac{14}{1} \cdot \frac{28}{1}$
$A = 196 \text{ feet}^2$

$980 + 196 = 1176 \text{ feet}^2$

31.
$$A = \pi r^2$$
$$A \approx 3.14(3^2)$$
$$A \approx 3.14(9)$$
$$A \approx 28.26$$
$$\frac{1}{2}A \approx \frac{1}{2}(28.26)$$
$$\frac{1}{2}A \approx 14.13 \text{ inches}^2$$

$$A = \frac{1}{2} \cdot b \cdot h$$
$$A = \frac{1}{2} \cdot \frac{6}{1} \cdot \frac{8}{1}$$
$$A = 24 \text{ inches}^2$$

$$14.13 + 24 \approx 38.12 \text{ in}^2$$

32.
$$A = l \cdot w$$
$$A = 9 \cdot 15$$
$$A = 135 \text{ feet}^2$$

$$\frac{135 \text{ feet}^2}{9 \text{ feet}^2} = \frac{x \text{ yards}^2}{1 \text{ yard}^2}$$
$$9x = 135$$
$$\frac{9x}{9} = \frac{135}{9}$$
$$x = 15 \text{ yard}^2$$

33.
$$A = \pi r^2$$
$$A = 3.14(40^2)$$
$$A = 3.24(1600)$$
$$A = 5024 \text{cm}^2$$

Approximately 5024cm² of glass

34.
$$A = l \cdot w$$
$$A = 17 \cdot 9$$
$$A = 153 \text{ feet}^2$$
153 feet² of floor tile.

35.
$$A = l \cdot w$$
$$A = 12 \cdot 16$$
$$A = 192 \text{cm}^2$$
192cm² of glass

36.
$$30 \text{ inches} = 2.5 \text{ feet}$$
$$A = l \cdot w$$
$$A = 5(25)$$
$$A = 12.5 \text{ feet}^2$$

12.5 feet² of fabric

37.
Each page $A = l \cdot w$
$$A = 8\frac{1}{2} \cdot 11$$
$$A = 93.5 \text{ inches}^2$$

80 pages are 93.5(80) = 7,480 inches²
The notebook has 7,480 inches² of paper.

38.
Area of the circle $= \pi r^2$
$$\approx (3.14)10^2$$
$$\approx (3.14(100)$$
$$\approx 314 \text{ inches}^2$$

Area of half the square, the triangle

$$A = \frac{1}{2} \cdot b \cdot h$$
$$A = \frac{1}{2} \cdot \frac{10}{1} \cdot \frac{20}{1}$$
$$A = 100 \text{ feet}^2$$

Area of the square 100(2) = 200 feet²
314 − 200 ≈ 114 feet² the area of the shaded region.

39. Area of large circle = πr^2
     $\qquad\qquad\quad = (3.14)10^2$
     $\qquad\qquad\quad = (3.14)100$
     $\qquad\qquad\quad = 314$ inches^2

Area of each small circle $\quad = \pi r^2$
$\qquad\qquad\qquad\qquad\quad = 3.14(5^2)$
$\qquad\qquad\qquad\qquad\quad = 3.14(25)$
$\qquad\qquad\qquad\qquad\quad = 314$ in^2
Area of both small circles (78.5)2 =
157 inches^2

Area of shaded region is 314 – 157 =
157 inches^2.

40. Area of rectangle
     $= l \cdot w$
     $= 25(20)$
     $= 500$ inch^2

Area of triangle
$= \dfrac{1}{2} \cdot b \cdot h$

$= \dfrac{1}{2} \cdot \dfrac{8}{1} \cdot \dfrac{10}{1}$

$= 40$ inches^2

Area of circle
$= \pi r^2$
$\approx (3.14)(4^2)$
$\approx (3.14)(16)$
≈ 50.27 inches^2

Area of triangle and circle 40 + 50.27 ≈
90.27 inches^2.
Area of shaded region is 500 – 90.27 ≈
409.73 inches^2.

12.3 Cubic Measure and Applications

1.
$$\frac{162\,feet^3}{27\,feet^3} = \frac{x\,feet^3}{1\,yard^3}$$
$$27x = 162$$
$$\frac{27x}{27} = \frac{162}{27}$$
$$x = 6\,yards^3$$

2.
$$\frac{8.5\,yards^3}{1\,yard^3} = \frac{x\,feet^3}{27\,feet^3}$$
$$x = 27(8.5)$$
$$x = 229.5\,feet^3$$

3.
$$\frac{15{,}552\,inches^3}{1728\,inches^3} = \frac{x\,feet^3}{1\,foot^3}$$
$$1728x = 15{,}552$$
$$\frac{1728x}{1728} = \frac{15552}{1728}$$
$$x = 9\,feet^3$$

4.
$$\frac{3\,feet^3}{1\,foot^3} = \frac{x}{1728\,inches^3}$$
$$x = 1728(3)$$
$$x = 5184\,inches^3$$

5.
$$\frac{93.312\,inches^3}{1728\,inches^3} = \frac{x\,feet^3}{1\,foot^3}$$
$$1728x = 93312$$
$$\frac{1728x}{1728} = \frac{93312}{1728}$$
$$x = 54\,feet^3$$

$$\frac{54\,feet^3}{27\,feet^3} = \frac{x\,yards^3}{1\,yard^3}$$
$$27x = 54$$
$$\frac{27x}{27} = \frac{54}{27}$$
$$x = 2\,yards^3$$

6. $2.2 \text{cm}^3 = 2200 \text{ mm}^3$

7. $93\text{m}^3 = 93{,}000{,}000 \text{ cm}^3$

8. $670\text{cm}^3 = 0.00067 \text{ m}^3$

9. $1300\text{mm}^3 = 0.0000013 \text{ m}^3$

10. $5200\text{cm}^3 = 0.0052 \text{ m}^3$

11. $V = l \cdot w \cdot h$
$V = 12 \cdot 10 \cdot 5$
$V = 600 \text{ cm}^3$

12. $V = l \cdot w \cdot h$
$V = 4 \cdot 9 \cdot 3$
$V = 108 \text{ inches}^3$

13. $V = \pi r^2 h$
$V \approx (3.14)(4^2)(8)$
$V \approx (3.14)(16)(8)$
$V \approx 401.92 \text{cm}^3$

14. $V = \pi r^2 h$
$V \approx (3.14)(10^2)(13)$
$V \approx (3.14)(100)(13)$
$V \approx 4082 \text{ inches}^3$

15. $V = \dfrac{4}{3}\pi r^3$
$V \approx \dfrac{4}{3}(3.14)(4^3)$
$V \approx \dfrac{4}{3}(3.14)(64)$
$V \approx 267.94\overline{6}$
$V \approx 267.95 \text{ feet}^3$

16. $V = \dfrac{4}{3}\pi r^3$
$V \approx \dfrac{4}{3}(3.14)(5^3)$
$V \approx \dfrac{4}{3}(3.14)(125)$
$V \approx 523.\overline{3} \qquad 523\dfrac{1}{3} \text{ cm}^3$

17. $V = s^3$
$V = 6^3$
$V = 216 \text{ cm}^3$

18. $V = s^3$
$V = 7^3$
$V = 343 \text{ inches}^3$

19.   3 feet = 36 inches
$V = \pi r^2 h$
$V \approx (3.14)(8^2)(36)$
$V \approx (3.14)(64)(36)$
$V \approx 7234.56 \text{ inches}^3$

20.   2m = 200cm
$V = l \cdot w \cdot h$
$V = 9 \cdot 7 \cdot 200$
$V = 12{,}600 \text{ cm}^3$

21.   4 feet = 48 inches
$V = \pi r^2 h$
$V \approx (3.14)(20^2)(48)$
$V \approx (3.14)(400)(48)$
$V \approx 60{,}319 \text{ inches}^3$
Volume of the water heater is
60,319 inches³

210

22.   8 inches $= \dfrac{8}{36} = \dfrac{2}{9}$ yard

Volume of mulch $= l \cdot w \cdot h$

$$V = 10 \cdot 12 \cdot \dfrac{2}{9}$$

$$V = \dfrac{80}{3} = 26.\overline{6} \text{ yard}^3$$

$26.\overline{6}$ or $26\dfrac{2}{3}$ yards^3 of mulch will be needed.

23.   4 inches $= \dfrac{4}{12} = \dfrac{1}{3}$ foot

Volume of cement $= l \cdot w \cdot h$

$$V = 12 \cdot 25 \cdot \dfrac{1}{3}$$

$$V = 100 \text{ feet}^3$$

$$\dfrac{100 \text{ feet}^3}{27 \text{ feet}^3} = \dfrac{x \text{ yards}^3}{1 \text{ yard}^3}$$

$$27x = 100$$

$$\dfrac{27x}{27} = \dfrac{100}{27}$$

$$x = 3.\overline{703} \text{ yards}^3$$

Approximately $3.\overline{703}$ yards^3 of cement will be needed.

24.   4 feet $= 1\dfrac{2}{3}$ yard

6 inches $= \dfrac{6}{36} = \dfrac{1}{6}$ yard

Volume of gravel $= l \cdot w \cdot h$

$$V = 15 \cdot 1\dfrac{1}{3} \cdot \dfrac{1}{6}$$

$$V = 3\dfrac{1}{3} \text{ yards}^3$$

$3\dfrac{1}{3}$ yards^3 of gravel will be needed.

25.   Volume of a sphere $= \dfrac{4}{3}\pi r^3$

$$V \approx \dfrac{4}{3}(3.14)5^3$$

$$V \approx \dfrac{4}{3}(3.14)125$$

$$V \approx 523.\overline{6}$$

There are about 523.6 inches^3 of air in the ball.

REVIEW

1. $$\dfrac{3 \text{ miles}}{1 \text{ mile}} = \dfrac{x \text{ feet}}{5280 \text{ feet}}$$

$$x = 5280(3)$$

$$x = 15840 \text{ feet}$$

2. $$\dfrac{13 \text{ yards}}{1 \text{ yard}} = \dfrac{x \text{ feet}}{3 \text{ feet}}$$

$$x = 3(13)$$

$$x = 39 \text{ feet}$$

$$\dfrac{39 \text{ feet}}{1 \text{ foot}} = \dfrac{x \text{ inches}}{12 \text{ inches}}$$

$$x = 39(12)$$

$$x = 468 \text{ inches}$$

3. $$\dfrac{84 \text{ inches}}{12 \text{ inches}} = \dfrac{x \text{ feet}}{1 \text{ feet}}$$

$$12x = 84$$

$$\dfrac{12x}{12} = \dfrac{84}{12}$$

$$x = 7 \text{ feet}$$

4. $$\dfrac{10 \text{ yard}}{1 \text{ yard}} = \dfrac{x \text{ feet}}{3 \text{ feet}}$$

$$x = 3(10)$$

$$x = 30 \text{ feet}$$

5.

$$\frac{48 \text{ feet}}{3 \text{ feet}} = \frac{x \text{ yards}}{1 \text{ yard}}$$
$$3x = 48$$
$$\frac{3x}{3} = \frac{48}{3}$$
$$x = 16 \text{ yards}$$

6.

$$\frac{54 \text{ inches}}{12 \text{ inches}} = \frac{x \text{ feet}}{1 \text{ foot}}$$
$$12x = 54$$
$$\frac{12x}{12} = \frac{54}{12}$$
$$x = 4.5 \text{ feet}$$

$$\frac{4.5 \text{ feet}}{3 \text{ feet}} = \frac{x \text{ yard}}{1 \text{ yard}}$$
$$3x = 4.5$$
$$\frac{3x}{3} = \frac{4.5}{3}$$
$$x = 1.5 \text{ yards}$$

7.

$$c^2 = a^2 + b^2$$
$$c^2 = 4^2 + 3^2$$
$$c^2 = 16 + 9$$
$$c^2 = 25$$
$$c = \sqrt{25}$$
$$c = 5 \text{ inches}$$

8.

$$c^2 = a^2 + b^2$$
$$c^2 = 7^2 + 7^2$$
$$c^2 = 49 + 49$$
$$c^2 = 98$$
$$c = \sqrt{98}$$
$$c \approx 9.90 \text{ cm}$$

9.

$$c^2 = a^2 + b^2$$
$$13^2 = a^2 + 5^2$$
$$169 = a^2 + 25$$
$$-25 = \qquad -25$$
$$144 = a^2$$
$$\sqrt{144} = a$$
$$12 = a$$
$$a = 12 \text{ feet}$$

10.

$$c^2 = a^2 + b^2$$
$$10^2 = a^2 + 8^2$$
$$100 = a^2 + 64$$
$$-64 = \qquad -64$$
$$36 = a^2$$
$$\sqrt{36} = a$$
$$6 = a$$
$$a = 6 \text{ mm}$$

11.

$$c^2 = a^2 + b^2$$
$$18^2 = a^2 + 13^2$$
$$324 = a^2 + 169$$
$$-169 = \qquad -169$$
$$155 = a^2$$
$$\sqrt{155} = a$$
$$12.45 \approx a$$
$$a \approx 12.45 \text{cm}$$

12.

$$c^2 = a^2 + b^2$$
$$45^2 = 27^2 + b^2$$
$$2025 = 729 + b^2$$
$$-729 = -729$$
$$1296 = b^2$$
$$\sqrt{1296} = b$$
$$36 = b$$
$$b = 36 \text{ inches}$$

13. $c^2 = a^2 + b^2$
     $c^2 = 2^2 + 3^2$
     $c^2 = 4 + 9$
     $c^2 = 13$
     $c = \sqrt{13}$
     $c \approx 3.61$ feet

14. $c^2 = a^2 + b^2$
     $c^2 = 2.5^2 + 3.6^2$
     $c^2 = 6.25 + 12.96$
     $c^2 = 19.21$
     $c = \sqrt{19.21}$
     $c \approx 4.38$ mm

15. $p = a + b + c$
     $p = 10 + 9 + 5$
     $p = 24$ inches

16. $p = a + b + c$
     $p = 8 + 8 + 6$
     $p = 22$ cm

17. $p = 4s$
     $p = 4(12)$
     $p = 48$ mm

18. $p = 4s$
     $p = 4(8)$
     $p = 32$ feet

19. $p = 2l + 2w$
     $p = 2(12) + 2(6)$
     $p = 24 + 12$
     $p = 36$ m

20. 4 feet = 48 inches
     $p = 2l + 2w$
     $p = 2(48) + 2(8)$
     $p = 96 + 16$
     $p = 112$ inches

21. $c = \pi d$
     $c \approx (3.14)(8)$
     $c \approx 25.12$ inches

22. $c = 2\pi r$
     $c \approx 2(3.14)2.5$
     $c \approx 15.7$ cm

23. $c = \pi d$
     $c \approx (3.14)(3.5)$
     $c \approx 10.99$ feet

24. $c = 2\pi r$
     $c \approx 2(3.14)19$
     $c \approx 119.32$ mm

25. $c = 2\pi r$
     $c \approx 2(3.14)22$
     $c \approx 138.16$ yards

26.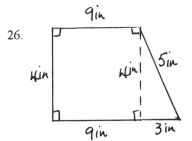

In the triangle:

$c^2 = a^2 + b^2$
$c^2 = 4^2 + 3^2$
$c^2 = 16 + 9$
$c^2 = 25$
$c = \sqrt{25}$
$c = 5$

$P = 4 + 9 + 5 + 12$
$P = 30$ inches

213

27.

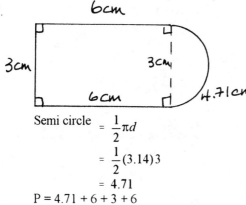

6cm

3cm   3cm

6cm   4.71cm

Semi circle $= \frac{1}{2}\pi d$

$= \frac{1}{2}(3.14)3$

$= 4.71$

P = 4.71 + 6 + 3 + 6

P = 19.71cm

28.

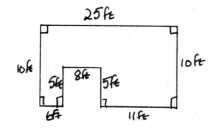

25ft

10ft   10ft

5ft 8ft 5ft

6ft   11ft

P = 25 + 10 + 11 + 5 + 8 + 5 + 6 + 10

P = 80 feet.

29.

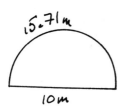

15.71m

10m

Circumference of the semi circle

$= \frac{1}{2}\pi d$

$\approx \frac{1}{2}(3.14)10$

≈ 15.7

P $\approx$ 15.7 + 10 $\approx$ 25.7m

30.

$p = 2l + 2w$
$p = 2(18) + 2(16)$
$p = 36 + 32$
$p = 68$ inches

$$\frac{68 \text{ inches}}{12 \text{ inches}} = \frac{x \text{ feet}}{1 \text{ foot}}$$
$$12x = 68$$
$$\frac{12x}{12} = \frac{68}{12}$$
$$x = 5.\overline{6} = 5\frac{2}{3} \text{ feet}$$

He will need $5\frac{2}{3}$ feet of picture frame.

31.

$p = 2l + 2w$
$p = 2(8) + 2(3)$
$p = 16 + 6$
$p = 22$ feet

$$\frac{22 \text{ feet}}{3 \text{ feet}} = \frac{x \text{ yards}}{1 \text{ yard}}$$
$$3x = 22$$
$$\frac{3x}{3} = \frac{22}{3}$$
$$x = 7\frac{1}{3} \text{ yards}$$

$7\frac{1}{3}$ yards of Halloween decorations

will be needed.

32.

$p = 4s$
$p = 4(25)$
$p = 100$ feet
He will need 100 feet of fence.

214

33.
$$p = 2l + 2w$$
$$p = 2(30) + 2(17)$$
$$p = 60 + 34$$
$$p = 94 \text{ feet}$$

94 feet of border will be needed.

34.
$$\frac{216 \text{ inches}^2}{144 \text{ inches}^2} = \frac{x \text{ feet}^2}{1 \text{ foot}^2}$$
$$144x = 216$$
$$\frac{144x}{144} = \frac{216}{144}$$
$$x = 1.5 \text{ feet}^2$$

35.
$$\frac{3 \text{ yards}^2}{1 \text{ yard}^2} = \frac{x \text{ yards}^2}{9 \text{ feet}^2}$$
$$x = 3(9)$$
$$x = 27 \text{ feet}^2$$

36.
$$\frac{3.5 \text{ feet}^2}{1 \text{ foot}^2} = \frac{x \text{ inches}^2}{144 \text{ inches}^2}$$
$$x = 3.5(144)$$
$$x = 504 \text{ inches}^2$$

37.
$$\frac{2.5 \text{ feet}^2}{1 \text{ foot}^2} = \frac{x \text{ inches}^2}{144 \text{ inches}^2}$$
$$x = 2.5(144)$$
$$x = 360 \text{ inches}^2$$

38.
$$\frac{\frac{1}{2} \text{ yard}^2}{1 \text{ yard}^2} = \frac{x \text{ feet}^2}{9 \text{ feet}^2}$$
$$x = 9\left(\frac{1}{2}\right)$$
$$x = 4\frac{1}{2} \text{ feet}^2$$

$$\frac{4\frac{1}{2} \text{ feet}^2}{1 \text{ foot}^2} = \frac{x \text{ inches}^2}{9 \text{ feet}^2}$$
$$x = 4\frac{1}{2}(144)$$
$$x = 648 \text{ inches}^2$$

39. $400\text{mm}^2 = 4\text{cm}^2$

40. $4\text{m}^2 = 4,000,000\text{mm}^2$

41. $10\text{km}^2 = 10,000,000\text{m}^2$

42.
$$A = \frac{1}{2} \cdot b \cdot h$$
$$A = \frac{1}{2} \cdot \frac{4}{1} \cdot \frac{6}{1}$$
$$A = 12 \text{ inches}^2$$

43.
$$A = \frac{1}{2} \cdot b \cdot h$$
$$A = \frac{1}{2} \cdot \frac{5}{1} \cdot \frac{10}{1}$$
$$A = 25\text{cm}^2$$

44.
$$A = s^2$$
$$A = 12^2$$
$$A = 144\text{mm}^2$$

45. $A = s^2$
         $A = 3^2$
         $A = 9$ feet^2

46.
         1m = 100cm
         $A = l \cdot w$
         $A = 100 \cdot 4$
         $A = 400$cm^2

47. $A = l \cdot w$
         $A = 17 \cdot 8$
         $A = 136$ inches^2

48. $A = b \cdot h$
         $A = 8 \cdot 4$
         $A = 32$ inches^2

49. $A = b \cdot h$
         $A = 13 \cdot 7$
         $A = 91$cm^2

50. $A = \pi r^2$
         $A \approx 3.14(8^2)$
         $A \approx 3.14(64)$
         $A \approx 200.96$ feet^2

51. $A = \pi r^2$
         $A \approx 3.14(12^2)$
         $A \approx 3.14(144)$
         $A \approx 452.16$m^2

52.

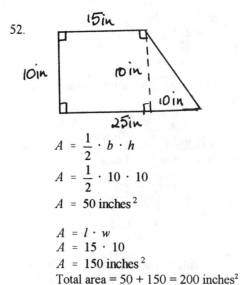

$A = \frac{1}{2} \cdot b \cdot h$

$A = \frac{1}{2} \cdot 10 \cdot 10$

$A = 50$ inches^2

$A = l \cdot w$
$A = 15 \cdot 10$
$A = 150$ inches^2
Total area = 50 + 150 = 200 inches^2

53.

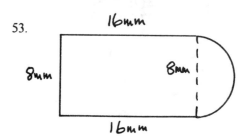

$A = l \cdot w$
$A = 16 \cdot 8$
$A = 128$mm^2

Area of semi-circle

$= \frac{1}{2}\pi r^2$

$\approx (0.5)(3.14)(4^2)$
$\approx (0.5)(3.14)(16)$
≈ 25.12mm^2

Total area is 128 + 25.12 = 153.12mm^2

54. Area of a semi circle

$$= \frac{1}{2}\pi r^2$$
$$\approx (0.5)(3.14)(9.5^2)$$
$$\approx (0.5)(3.14)(90.25)$$
$$\approx 141.6925 \text{m}^2$$

55.

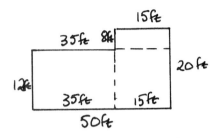

$$A = l \cdot w$$
$$A = 15 \cdot 20$$
$$A = 300 \text{ feet}^2$$

$$A = l \cdot w$$
$$A = 35 \cdot 12$$
$$A = 420 \text{ feet}^2$$
Total area is $300 + 420 = 720 \text{ feet}^2$

56. $$A = l \cdot w$$
$$A = 25(7.5)$$
$$A = 187.5 \text{ feet}^2$$

$$\frac{9 \text{ feet}^2}{1 \text{ yard}^2} = \frac{187.5 \text{ feet}^2}{x \text{ yards}^2}$$
$$9x = 187.5$$
$$\frac{9x}{9} = \frac{187.5}{9}$$
$$x = 20.8\overline{3} \text{ yards}^2$$
$20.8\overline{3}$ yards^2 of carpet.

57. $$A = l \cdot w$$
$$A = (9.5)(12)$$
$$A = 114 \text{cm}^2$$
114cm^2 of glass

58. $$A = l \cdot w$$
$$A = (8.5)(12.5)$$
$$A = 106.25 \text{ feet}^2$$
106.25 feet^2 of paneling

59. Area of the room $= l \cdot w$
$$= 8 \cdot 7\frac{1}{2}$$
$$= \frac{8}{1} \cdot \frac{15}{2}$$
$$= 60 \text{m}^2$$

Area of carpet $= l \cdot w$
$$= 6 \cdot 5\frac{1}{2}$$
$$= \frac{6}{1} \cdot \frac{11}{2}$$
$$= 33 \text{m}^2$$

The uncarpeted area is $60 - 33 = 27 \text{m}^2$.

60. $$\frac{27 \text{ feet}^3}{1 \text{ yard}^3} = \frac{243 \text{ feet}^3}{x \text{ yards}}$$
$$27x = 243$$
$$\frac{27x}{27} = \frac{243}{27}$$
$$x = 9 \text{ yards}^3$$

61. $10\text{m}^3 = 10,000,000\text{cm}^3$

62. $127\text{mm}^3 = 0.127\text{cm}^3$

63.

$$\frac{1728 \text{ inches}^3}{1 \text{ foot}^3} = \frac{4320 \text{ inches}^3}{x \text{ feet}^3}$$
$$1728x = 4320$$
$$\frac{1728x}{1728} = \frac{4320}{1728}$$
$$x = 2.5 \text{ feet}^3$$

64.

$$\frac{27 \text{ feet}^3}{1 \text{ yard}^3} = \frac{108 \text{ feet}^3}{x \text{ yards}^3}$$
$$27x = 108$$
$$\frac{27x}{27} = \frac{108}{27}$$
$$x = 4 \text{ yards}^3$$

65.

$$V = s^3$$
$$V = 16^3$$
$$V = 4096 \text{ inches}^3$$

66.

$$V = \frac{4}{3}\pi r^3$$
$$V \approx \frac{4}{3}(3.14)(3^3)$$
$$V \approx \frac{4}{3}(3.14)(27)$$
$$V \approx 113.04 \text{cm}^3$$

67.

$$V = \pi r^2 h$$
$$V \approx (3.14)(10^2)(27)$$
$$V \approx (3.14)(100)(27)$$
$$V \approx 8478 \text{inches}^3$$

68.

$$7 \text{ inches} = \frac{7}{12} \text{ foot}$$
$$V = l \cdot w \cdot h$$
$$V = 8 \cdot 3 \cdot \frac{7}{12}$$
$$V = 14 \text{ feet}^3$$

69.

$$4 \text{ inches} = \frac{4}{12} = \frac{1}{3} \text{ foot}$$
$$V = l \cdot w \cdot h$$
$$V = 90 \cdot 3 \cdot \frac{2}{3}$$
$$V = 90 \text{ feet}^3$$

$$\frac{90 \text{ feet}^3}{x \text{ yards}^3} = \frac{27 \text{ feet}^3}{1 \text{ yard}^3}$$
$$27x = 90$$
$$\frac{27x}{27} = \frac{90}{27}$$
$$x = 3\frac{1}{3} \text{ yards}^3$$

$3\frac{1}{3}$ yards^3 of cement

70.

$$30 \text{ inches} = 2.5 \text{ feet}$$
$$V = \pi r^2 h$$
$$V \approx (3.14)(2.5^2)(5)$$
$$V \approx (3.14)(6.25)(5)$$
$$V \approx 98.125 \text{ feet}^3$$

The water tank has a volume of 98.125 feet^3.

71.

$$V = \pi r^2 h$$
$$V \approx (3.14)(3^2)(10)$$
$$V \approx (3.14)(9)(10)$$
$$V \approx 282.6 \text{cm}^3$$

The can has a volume of 70.65cm^3.

72.
$$V = l \cdot w \cdot h$$
$$V = 10 \cdot 6 \cdot 4$$
$$V = 240\text{cm}^3$$

$$V = s^3$$
$$V = 2^3$$
$$V = 8\text{cm}^3$$

$240 \div 8 = 30$
The box has a volume of 240cm³ and will hold 30 cubes.

PRACTICE TEST

1.
$$\frac{50 \text{ inches}}{x \text{ feet}} = \frac{12 \text{ inches}}{1 \text{ foot}}$$
$$12x = 50$$
$$\frac{12x}{12} = \frac{50}{12}$$
$$x = 4.1\overline{6} \text{ feet}$$

2.
$$\frac{1 \text{ mile}}{5280 \text{ feet}} = \frac{3 \text{ miles}}{x \text{ feet}}$$
$$x = 15,840 \text{ feet}$$

$$\frac{3 \text{ feet}}{1 \text{ yard}} = \frac{15840 \text{ feet}}{x \text{ yards}}$$
$$3x = 15840$$
$$\frac{3x}{3} = \frac{15840}{3}$$
$$x = 5280 \text{ yards}$$

3.   382cm = 3.82m

4.   9.5km = 9500m

5.
$$\frac{1 \text{ foot}}{12 \text{ inches}} = \frac{8.5 \text{ feet}}{x \text{ inches}}$$
$$x = (8.5)12$$
$$z = 102 \text{ inches}$$

The ceiling is 102 inches high.

6.
$$\frac{36 \text{ inches}}{1 \text{ yard}} = \frac{53 \text{ inches}}{x}$$
$$36x = 53$$
$$\frac{36x}{36} = \frac{53}{36}$$
$$x = 1.47\overline{2} \text{ yards}$$

The door is 1.47$\overline{2}$ yards wide.

7.   390cm² = 0.039m²

8.
$$c^2 = a^2 + b^2$$
$$c^2 = 1^2 + 2^2$$
$$c^2 = 1 + 4$$
$$c^2 = 5$$
$$c = \sqrt{5}$$
$$c \approx 2.24 \text{ m}$$

9.
$$c^2 = a^2 + b^2$$
$$20^2 = a^2 + 16^2$$
$$400 = a^2 + 256$$
$$-256 = \qquad -256$$
$$144 = a^2$$
$$\sqrt{144} = a$$
$$12 \text{ feet} = a$$

10.
$$c^2 = a^2 + b^2$$
$$13^2 = a^2 + 5^2$$
$$169 = a^2 + 25$$
$$-25 = \qquad -25$$
$$144 = a^2$$
$$\sqrt{144} = a$$
$$12 \text{ inches} = a$$

219

11.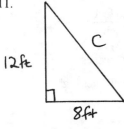

12ft

8ft

$$c^2 = a^2 + b^2$$
$$c^2 = 12^2 + 8^2$$
$$c^2 = 144 + 64$$
$$c^2 = 208$$
$$c = \sqrt{208}$$
$$c = 14.42 \text{ feet}$$

12. $P = 4s$
     $P = 4(19)$
     $P = 76 \text{ feet}$

13. $P = 4s$
     $28 = 4s$
     $\dfrac{28}{4} = \dfrac{4s}{4}$
     $7\text{mm} = s$

14. $P = a + b + c$
     $P = 5.6 + 3.8 + 4.6$
     $P = 14\text{cm}$

15. $p = 2l \cdot 2w$
     $p = 2(72) + 2(36)$
     $p = 144 + 72$
     $p = 216\text{mm}$

16. $p = a + b + c + d$
     $p = 34 + 17 + 34 + 17$
     $p = 102 \text{ yards}$

17. Circumference of a semi circle
     $= \dfrac{1}{2}\pi d$
     $\approx (0.5)(3.14)(9)$
     ≈ 14.13
     $14.13 + 10 + 10 \approx 34.13 \text{ feet.}$

18. $c = 2\pi r$
     $c \approx 2(3.14)(2.5)$
     $c \approx 15.7 \text{ inches}$

19. $c = \pi d$
     $c \approx 3.14(7)$
     $c \approx 21.98 \text{ feet}$
     21.98 feet of decorative fence.

20. $\dfrac{720 \text{ inches}^2}{x \text{ feet}^2} = \dfrac{144 \text{ inches}^2}{1 \text{ foot}^2}$
     $144x = 720$
     $\dfrac{144x}{144} = \dfrac{720}{144}$
     $x = 5 \text{ feet}^2$

21. $200\text{mm} = 20\text{cm}$

22. $A = s^2$
     $A = 19^2$
     $A = 361 \text{ inches}^2$

23. $A = b \cdot h$
     $A = 17 \cdot 7$
     $A = 119\text{m}^2$

24. $A = \dfrac{1}{2} \cdot b \cdot h$
     $A = \dfrac{1}{2} \cdot 9 \cdot 17$
     $A = 76.5\text{cm}^2$

25.
$$3 \text{ yards} = 3(36) \text{ inches} = 108 \text{ inches}$$
$$A = l \cdot w$$
$$A = 108 \cdot 15$$
$$A = 1620 \text{ inches}^2$$

26.
$$A = \pi r^2$$
$$A \approx (3.14)(2.5^2)$$
$$A \approx (3.14)(6.25)$$
$$A \approx 19.625 \text{ inches}^2$$

27.
$$A = l \cdot w$$
$$A = 13 \cdot 9$$
$$A = 117 \text{ feet}^2$$

$$\frac{117 \text{ feet}^2}{x \text{ yards}^2} = \frac{9 \text{ feet}^2}{1 \text{ yard}^2}$$
$$9x = 117$$
$$\frac{9x}{9} = \frac{117}{9}$$
$$x = 13$$

13 yards² of carpet.

28.
$$V = \frac{4}{3} \pi r^3$$
$$V \approx \frac{4}{3}(3.14)(7^3)$$
$$V \approx \frac{4}{3}(3.14)(343)$$
$$V \approx 1436.03 \text{cm}^3$$

29.
$$V = s^3$$
$$V = (3.5)^3$$
$$V = 42.875 \text{ inches}^3$$

30.
$$3 \text{ feet} = 1 \text{ yard} \quad 5 \text{ inches} = \frac{5}{36} \text{ yard}$$
$$V = l \cdot w \cdot h$$
$$V = 75 \cdot 1 \cdot \frac{5}{36}$$
$$V = 10\frac{5}{12} \text{ yards}^3$$

$10\frac{5}{12}$ yards³ of cement

1. A <u>Ratio</u> is an indicated quotient.

2. A <u>Proportion</u> is an equation whose members are ratios.

3. The <u>Zero Product Property</u> states that $a \cdot 0 = 0$, where $a \in$ Reals.

4. The <u>Metric System</u> is the system of measurement used by most of the world.

5. The <u>Property of Reciprocals</u> states that, for all $a \in$ Reals, $a \neq 0$, $a \cdot \dfrac{1}{a} = 1$.

6. In a right triangle, the square of the length of the hypotenuse is equal to the sum of the squares of the lengths of the legs. This is called the <u>Pythagorean Theorem</u>.

7. The statement <u>the product of the means equals the product of the extremes</u> means that, for all a, b, c, $d \in$ Reals, $b \neq 0$, $d \neq 0$, if $\dfrac{a}{b} = \dfrac{c}{d}$, then $bc = ad$.

8. The <u>Distributive Property</u> states that, for all a, b, $c \in$ Reals, $a(b + c) = ab + ac$.

9. The <u>Property of Additive Inverses</u> states that, for all $a \in$ Reals, $a + (-a) = 0$.

10. The <u>Identity Property of Multiplication</u> states that, for all $a \in$ Reals, $a \cdot 1 = a$.

11. $\dfrac{1}{8} = 0.125 = 12.5\%$

12. $\dfrac{2}{3} = 0.\overline{6} = 66\dfrac{2}{3}\%$

13. $\dfrac{13}{25} = 0.52 = 52\%$

14. $2\dfrac{3}{10} = 2.3 = 230\%$

15. $\dfrac{1}{200} = 0.005 = \dfrac{1}{2}\%$

16. $\dfrac{3}{1000} = 0.003 = 0.3\%$

17. $18 : 45 \quad \dfrac{18}{45} = \dfrac{2}{5} = 2 : 5$

18. $\dfrac{1}{3} : \dfrac{1}{2}$

    $\dfrac{\frac{1}{3}}{\frac{1}{2}} = \dfrac{1}{3} \div \dfrac{1}{2} = \dfrac{1}{3} \cdot \dfrac{2}{1} =$

    $\dfrac{2}{3} = 2 : 3$

19. $\dfrac{0.6}{60} = \dfrac{6}{600} = \dfrac{1}{100}$

    $1 : 100$

222

20. 20 seconds to 7 minutes
    20 seconds to 420 seconds

    $\dfrac{20}{420} = \dfrac{1}{21}$

    $1:21$

21. 385 miles per 15 gallons of gas

    $\dfrac{385}{15} = 25\dfrac{2}{3}$ mpg

22. 2600 TVs to 1000 households

    $\dfrac{2600}{1000} = 2.6$ TVs per household

23. 60% of 152 $= x$
    $(0.6)\,152 = x$
    $91.2 = x$

24. $\dfrac{55}{100} = \dfrac{123}{x}$
    $55x = 12300$
    $\dfrac{55x}{55} = \dfrac{12300}{55}$
    $x = 223.6\overline{3}$

25. $\dfrac{x}{100} = \dfrac{63.96}{82}$
    $82x = 6396$
    $\dfrac{82x}{82} = \dfrac{6396}{82}$
    $x = 78\%$

26. 15.5% of $x = 45$
    $(0.155)x = 45$
    $0.155x = 45$
    $\dfrac{0.155x}{0.155} = \dfrac{45}{0.155}$
    $x = 290.32$

27. $\dfrac{3 \text{ feet}}{x \text{ inches}} = \dfrac{1 \text{ foot}}{12 \text{ inches}}$
    $x = 3\,(12)$
    $x = 36$ inches

28. $\dfrac{2 \text{ yards}^2}{x \text{ feet}^2} = \dfrac{1 \text{ yard}^2}{9 \text{ feet}^2}$
    $x = 2\,(9)$
    $x = 18$ feet^2

29. $\dfrac{81 \text{ feet}^3}{x \text{ yards}^3} = \dfrac{27 \text{ feet}^3}{1\,\dfrac{\text{yard}}{3}}$
    $27x = 81$
    $\dfrac{27x}{27} = \dfrac{81}{27}$
    $x = 3$ yards^3

30. $300\,\text{mm}^2 = 3\,\text{cm}^2$

31. $2.5\,\text{m} = 250\,\text{cm}$

32. $2500\,\text{m} = 2.5\,\text{km}$

33. $14 selling price, $8 purchase price

    $\dfrac{14}{8} = \dfrac{7}{4}$

    $7:4$

34. $10 profit, $50 selling price

    $\dfrac{10}{50} = \dfrac{1}{5}$

    $1:5$

35. $\dfrac{300 \text{ children}}{100 \text{ families}}$

$\dfrac{300}{100} = \dfrac{3}{1}$

3 children per family

36. $\dfrac{102}{100} = \dfrac{32}{x}$

$102x = 3200$

$\dfrac{102x}{102} = \dfrac{3200}{102}$

$x = 31.37$

The previous charge was $31.37.

37.
Jennie	$ 8.95
Max	$10.95
Total	$19.90

6% of $19.90 = x
(0.06)19.90 = 1.19

15% of $19.90 = x
(0.15)(19.9) = 2.99

Bill	$19.90
Tax	1.19
Tip	2.99
Total	$24.08

38. $c^2 = a^2 + b^2$

$c^2 = 6^2 + 8^2$

$c^2 = 36 + 64$

$c^2 = 100$

$c = \sqrt{100}$

$c = 10 \text{ inches}$

39. $c^2 = a^2 + b^2$

$13^2 = a^2 + 12^2$

$169 = a^2 + 144$

$-144 = -144$

$25 = a^2$

$\sqrt{25} = a$

$5\text{cm} = a$

40. $P = 2l \cdot 2w$

$20 = 2l \cdot 2(4)$

$20 = 2l + 8$

$-8 = -8$

$\dfrac{12}{2} = \dfrac{2l}{2}$

$6 \text{ feet} = l$

The length of the rectangle is 6 feet.

41. $A = \dfrac{1}{2} \cdot b \cdot h$

$A = \dfrac{1}{2} \cdot 9 \cdot 10$

$A = 45\text{m}^2$

42. $V = s^3$

$V = 2^3$

$V = 8 \text{ yards}^3$

43. $c = \pi d$

$c \approx (3.14)(4)$

$c \approx 12.56\text{cm}$

44.

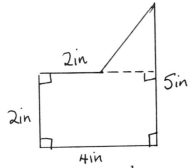

Area of triangle $= \frac{1}{2} \cdot b \cdot h2$

$= \frac{1}{2} \cdot 2 \cdot 3$

$= 3 \text{ inches}^2$

Area of rectangle $= l \cdot w$
$= 4 \cdot 2$
$= 8 \text{ inches}^2$

Total area $= 3 + 8 = 11 \text{ inches}^2$

CHAPTER 13: GRAPHS AND CHARTS

13.1 The Rectangular Coordinate Plane

1.- 8.

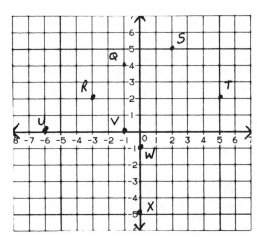

9. - 14.

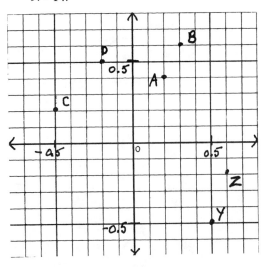

Scale: 10 blocks = 1 unit

15. I
16. IV
17. between III & IV
18. III
19. between II & III
20. II
21. (0,0)
22. (2,3)

23. (-9,0)
24. (6,0)
25. (4,-6)
26. (-2,-2)
27. (-4,7)
28. (0,9)

13.2 Graphing Linear Equations

1.-2.

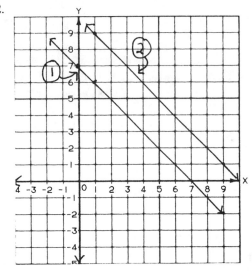

3.-4.

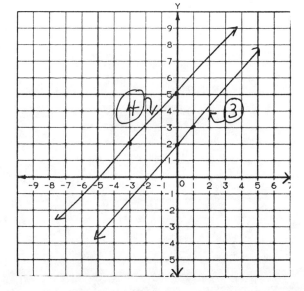

5.-6.

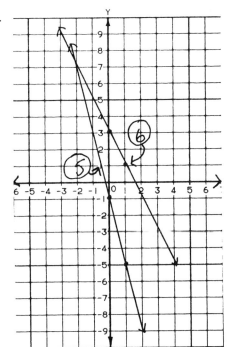

9.-10.

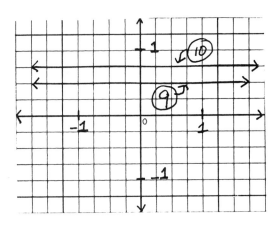

Scale: 4 blocks = 1 unit

7.-8.

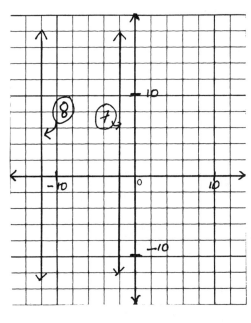

Scale: 1 block = 2 units

11.-12.

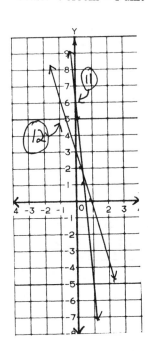

227

13.-14.

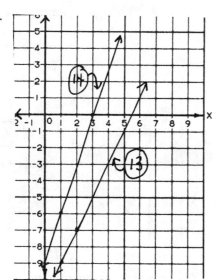

17.-18.

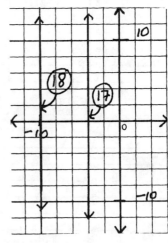

1 block = 2 units

15.-16.

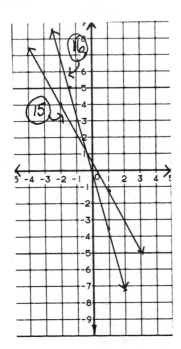

19.-20.

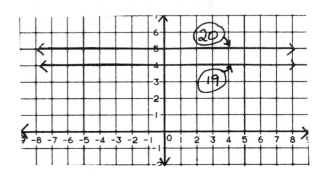

21. 0
22. 9
23. -9
24. -15
25. 2
26. -6
27. $3\frac{2}{3}$ or $\frac{11}{3}$
28. 3
29. $\frac{5}{3}$ or $1\frac{2}{3}$
30. $\frac{4}{3}$ or $1\frac{1}{3}$
31. $-4\frac{1}{2}$ or $-\frac{9}{4}$
32. -5

228

13.3 Interpreting Statistical Graphs and Charts

1. Answers may vary.
2. Answers may vary.
3. Answers may vary.
4. Answers may vary.
5. Answers may vary.
6. Answers may vary.
7. Answers may vary.

13.4 Constructing Statistical Graphs and Charts

1 a.

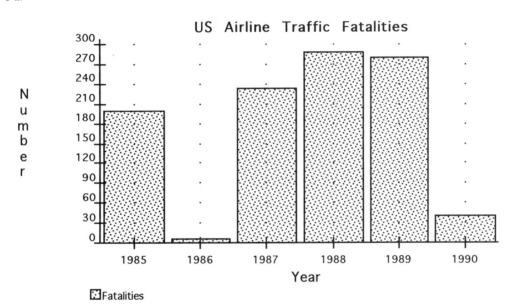

1. b.

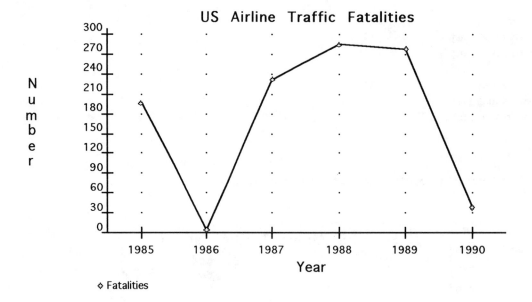

2. a.

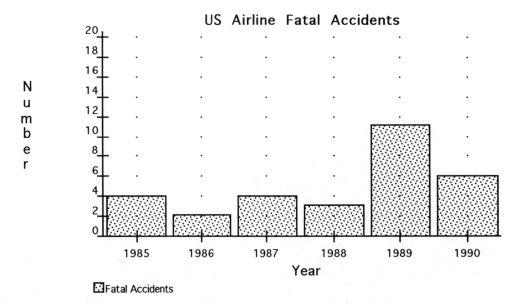

2. b.

US Airline Fatal Accidents

3. a.

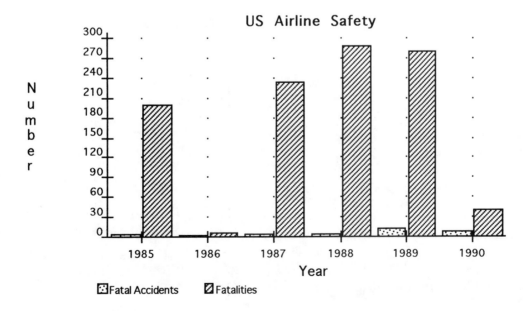

US Airline Safety

231

3. b.

4. a.

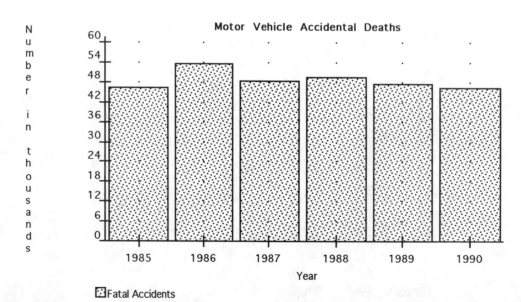

232

4. b.

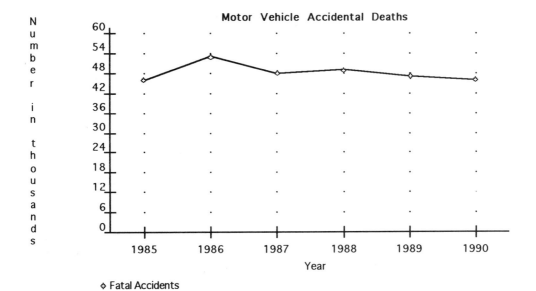

5. a.

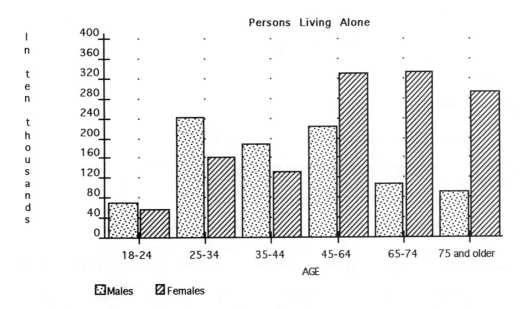

5. b.

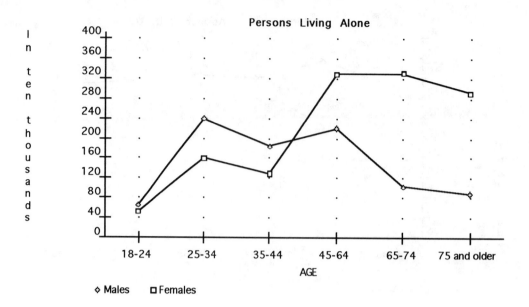

Persons Living Alone

In ten thousands

AGE

◇ Males   □ Females

6. a.

Age	Persons Living Alone[1]
18-24	1451
25-34	3973
35-44	3139
45-64	5503
65-74	4351
75 and older	4825

[1]In thousands

234

6. b.

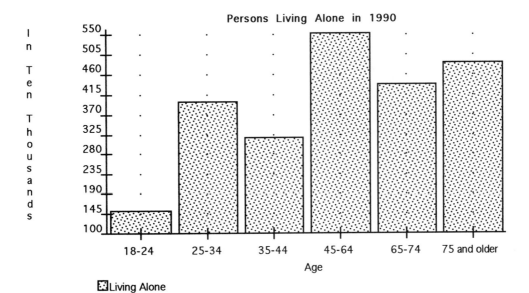

6. c.

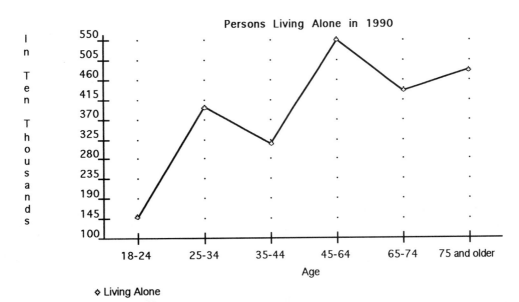

7.

Motor Vehicles in fatal accidents: 1991

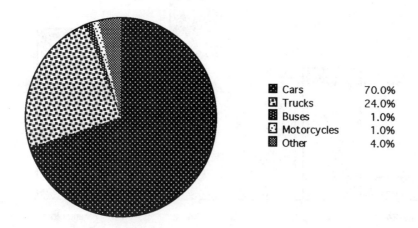

▨	Cars	70.0%
▣	Trucks	24.0%
▦	Buses	1.0%
▨	Motorcycles	1.0%
▨	Other	4.0%

Review

1. - 4.

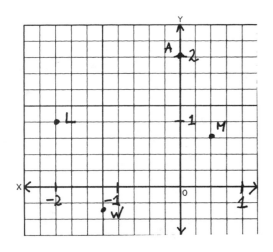

Scale: 4 blocks = 1 unit

5. (1,4)
6. (2, -2)
7. (0, 2)
8. (-5, -2)
9. IV
10. III
11. between II & III
12. I

13.-15.

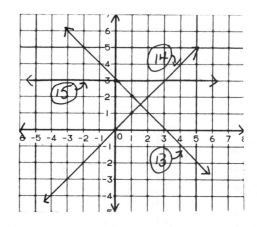

16.-18.

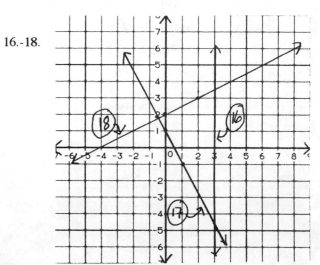

19. 4

20. 2

21. 3

22. -3

23. 1

24. -4

25. a. Public - 4 yr
    b. Private - 2 yr
    c. Public - 4 yr & Private - 2 yr
    d. Public - 2 yr & Private - 4 yr

26. a. Public - 4 yr & Public - 2 yr
    b. Private - 4 yr & Private - 2 yr
    c. Private - 4 yr
    d. Private - 2 yr

27. a. 1990
    b. 1960
    c. Since 1960 the number of marriages has been increasing.

28. a. ABC Fund
    b. From 1989 to 1990
    c. All As Fund
    d. ABC Fund $41,000;
       All As Fund $30,000

29. a. Basketball
    b. Tie: Skiing and Golf
    c. 15
    d. 21

30.

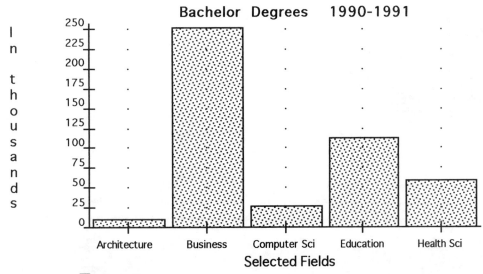

31.

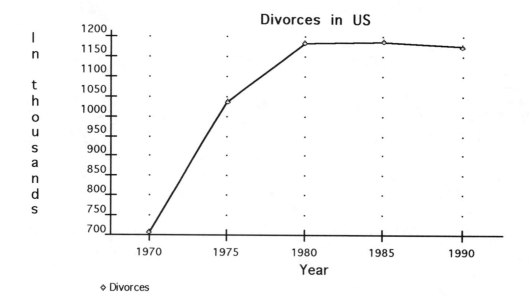

Divorces in US

In thousands

◇ Divorces

32.

Senior Pizza Preferences at IM U

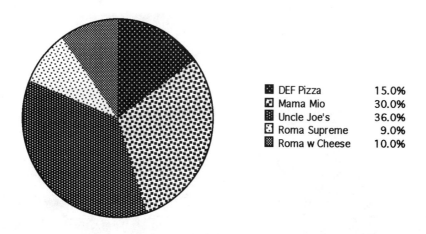

▨	DEF Pizza	15.0%
▣	Mama Mio	30.0%
▦	Uncle Joe's	36.0%
▫	Roma Supreme	9.0%
▨	Roma w Cheese	10.0%

Practice Test

1. II, (-9, 7)
2. IV, (3, -1)
3. I, (1, 9)
4. between II & III, (-4,0)
5. between III & IV, (0, -6)
6. III, (-8, -3)
7. I, (4, 3)

8.-13.

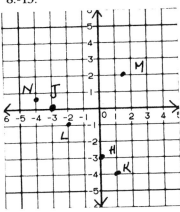

14.-16.

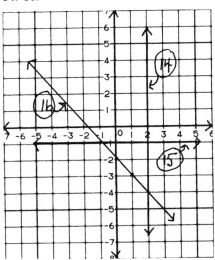

17.-18.

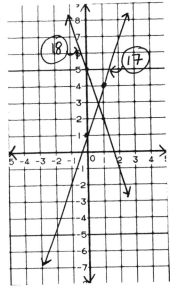

19. (-2, 0)
20. (0, -3)
21. (-4, 3)
22. (-1, -1.5)
23. a. Circle or Pie
    b. Freshmen, 350
    c. Juniors, 190
24. a. Line Graph
    b. Median age at first marriage is
       increasing for men.
       Median age at first marriage is
       increasing for women.
    c. The difference is narrowing.
25. a. Bar Graph
    b. The marriage rate decreased from
       1950 to 1960 and then increased
       from 1960 to 1970.
    c. The divorce rate decreased from
       1950 to 1960 and then increased
       from 1960 to 1970.
    d. No.

239

26.

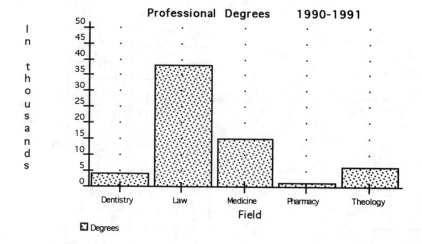

27.

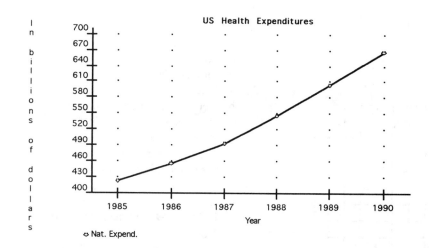

28.

A Freshman's Monthly Budget

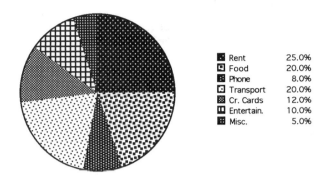

▣	Rent	25.0%
▣	Food	20.0%
▣	Phone	8.0%
▣	Transport	20.0%
▣	Cr. Cards	12.0%
▣	Entertain.	10.0%
▣	Misc.	5.0%

PRACTICE FINAL EXAM

1. The <u>Distributive Property</u> states that for all real numbers $a(b + c) = ab + ac$.

2. The <u>Cumulative Property of Multiplication</u> states that for all real numbers, $ab = ba$.

3. <u>Variables</u> are letters used to represent numbers.

4. A <u>Prime number</u> is a whole number greater than one having only two factors, namely, itself and 1.

5. The <u>Absolute value</u> of a number represents the distance that number is from zero on the number line.

6. The <u>Addition Property of Equality</u> states that for all real numbers, if $a = b$, then $a + c = b + c$.

7. The <u>Multiplication Property of Equality</u> states that for all real numbers, if $a = b$, then $ac = bc$.

8. The <u>Associative Property of Addition</u> states that for all real numbers, $a + (b + c) = (a + b) + c$.

9. A <u>Proportion</u> is an equation whose members are ratios.

10. In the expression $2y + 64$, the number 64 is a <u>Constant</u>.

11. $0.017 = \dfrac{17}{1000}$

12.
$$\begin{array}{r} 21.680 \\ 0.563 \\ +\underline{1.400} \\ 23.643 \end{array}$$

13.
$$\begin{array}{r} 5.060 \\ -\underline{4.982} \\ 0.078 \end{array}$$

14.
$$\begin{array}{r} 30.4 \\ 0.04\overline{)1.216} \\ \underline{12} \\ 016 \\ 16 \end{array}$$

15. $(0.004)(0.05)$
$4 \cdot 5 = 20$
$0.00020 = 0.0002$

16. $8.7\% = 0.087$

17. $18\dfrac{2}{3} = \dfrac{56}{3}$

18.
$$\begin{array}{r} 4\dfrac{5}{6} = 3\dfrac{44}{24} \\ -2\dfrac{7}{8} = -2\dfrac{21}{24} \\ 1\dfrac{23}{24} \end{array}$$

19.
$$\dfrac{3}{5} > \dfrac{5}{9}$$
$$\dfrac{27}{45} > \dfrac{25}{45}$$

242

20. $$\frac{-15 + 8 + (-16) + 3}{4} =$$
$$\frac{-20}{4} = -5$$
The average is -5

21. 78,907 rounded to the nearest ten is 78,910.

22. $$\frac{1\frac{2}{3}}{1\frac{1}{2}} = \frac{\frac{5}{3}}{\frac{3}{2}} =$$
$$\frac{5}{3} \div \frac{3}{2} = \frac{5}{3} \cdot \frac{2}{3} = \frac{10}{9} = 1\frac{1}{9}$$

23.

3	315
3	105
5	35
7	7
	1

$$315 = 3 \cdot 3 \cdot 5 \cdot 7$$

24. $$24\% = \frac{24}{100} = \frac{6}{25}$$

25. $$12 - 6 \div 3 = 12 - 2 = 10$$

26. $$\frac{2\frac{2}{3}}{4\frac{4}{5}} = \frac{\frac{8}{3}}{\frac{24}{5}} = \frac{8}{3} \cdot \frac{5}{24}$$
$$= \frac{40}{72} = \frac{5}{9}$$

27. $$\frac{1.8}{6.4} = \frac{x}{0.16}$$
$$6.4x = (1.8)(0.16)$$
$$6.4x = 0.288$$
$$\frac{6.4x}{6.4} = \frac{0.288}{6.4}$$
$$x = 0.045$$

28. $$12 + 3(6)^2 = 12 + 3(36) =$$
$$12 + 108 = 120$$

29. $$9\frac{2}{5} = \frac{47}{5}$$
Reciprocal is $\frac{5}{47}$

30. 897.00984 rounded to the nearest hundredth is 897.01

31. $$\frac{1}{2}x - \frac{1}{4} = 2$$
$$\cancel{4}^2(\frac{1}{\cancel{2}}x) - \cancel{4}(\frac{1}{\cancel{4}}) = 4 \cdot 2$$
$$2x - 1 = 8$$
$$1 = 1$$
$$2x = 9$$
$$\frac{2x}{2} = \frac{9}{2}$$
$$x = \frac{9}{2} = 4\frac{1}{2}$$

32. $$5 - 3y = 8 - y$$
$$3y = 3y$$
$$5 = 8 + 2y$$
$$-8 = -8$$
$$-3 = 2y$$
$$-\frac{3}{2} = \frac{2y}{2}$$
$$-\frac{3}{2} = y$$
$$y = -\frac{3}{2} = -1\frac{1}{2}$$

33.
$$20 = 8\% \text{ of } x$$
$$20 = (0.08)\,x$$
$$\frac{20}{0.08} = \frac{0.08x}{0.08}$$
$$250 = x$$
$$x = 250$$

34.
$$5\frac{1}{4}\% \text{ of } \$860 = x$$
$$(0.0525)\,860 = x$$
$$45.15 = x$$

The tax is \$45.15

35.
$$(-1)(3)(-2)(-5)(-2) = 60$$

36.
$$(-15) \div (-3) = 5$$

37.
$$|-9| + |-3| + |0| + |9| =$$
$$9 + 3 + 0 + 9 = 21$$

38.
$$2b^2 - 5b - 3, \text{ for } b = -2$$
$$2(b^\bullet) - 5(b) - 3 =$$
$$2(-2)^2 - 5(-2) - 3 =$$
$$8 + 10 - 3 = 15$$

39.
$$\frac{x}{100} = \frac{2.5}{4.00}$$
$$4x = 250$$
$$\frac{4x}{4} = \frac{250}{4}$$
$$x = 62.5\%$$

40.
$$p = 2l \cdot 2w$$
$$p = 2(5.3) + 2(2.7)$$
$$p = 10.6 + 5.4$$
$$p = 16 \text{ inches}$$

$$A = l \cdot w$$
$$A = (5.3)(2.7)$$
$$A = 14.31 \text{ inches}^2$$

41.
$$c^2 = a^2 + b^2$$
$$c^2 = 5^2 + 12^2$$
$$c^2 = 25 + 144$$
$$c^2 = 169$$
$$c = \sqrt{169}$$
$$c = 13$$

$$P = a \cdot b \cdot c$$
$$P = 5 \cdot 12 \cdot 13$$
$$P = 30\text{cm}$$

$$A = \frac{1}{2} \cdot b \cdot h$$
$$A = \frac{1}{2} \cdot 5 \cdot 12$$
$$A = 30\text{cm}^2$$

42.
$$A = \frac{1}{2} \cdot b \cdot h$$
$$A = \frac{1}{2} \cdot 7 \cdot 3$$
$$A = \frac{21}{2} = 10.5 \text{ feet}^2$$

43.
$$A = \pi r^2$$
$$50.24 \approx 3.14 r^2$$
$$\frac{50.24}{3.14} \approx \frac{3.14 r^2}{3.14}$$
$$16 \approx r^2$$
$$\sqrt{16} \approx r$$
$$4 \approx r$$

The radius is 4 inches.

44.
$$V = s^3$$
$$V = 2^3$$
$$V = 8 \text{ inches}^3$$

45.
$$\frac{0.05}{1.2} = \frac{x}{6}$$
$$1.2x = (0.05)6$$
$$1.2x = 0.30$$
$$\frac{1.2x}{1.2} = \frac{0.3}{1.2}$$
$$x = 0.25$$

46.
$$\frac{y}{3} + \frac{2}{7} = -\frac{1}{2}$$
$$42(\frac{y}{3}) + 42(\frac{2}{7}) = 42(-\frac{1}{2})$$
$$14y + 12 = -21$$
$$-12 = -12$$
$$14y = -33$$
$$\frac{14y}{14} = \frac{-33}{14}$$
$$y = \frac{-33}{14} = -2\frac{5}{14}$$

47.
$$8(x - \frac{1}{2}) = 2(x - \frac{1}{4})$$
$$8x - 4 = 2x - \frac{1}{2}$$
$$2(8x) - 2(4) = 2(2x) - 2(\frac{1}{2})$$
$$16x - 8 = 4x - 1$$
$$-4x = -4x$$
$$12x - 8 = -1$$
$$8 = 8$$
$$12x = 7$$
$$\frac{12x}{12} = \frac{7}{12}$$
$$x = \frac{7}{12}$$

48.
$$1.2\% = \frac{1.2}{100} = \frac{12}{1000} = \frac{3}{250}$$

49.
$$c^2 = a^2 + b^2$$
$$20^2 = 16^2 + b^2$$
$$400 = 256 + b^2$$
$$-256 = -256$$
$$144 = b^2$$
$$\sqrt{144} = b$$
$$12 = b$$
Side b is 12 inches long.

50.
$$\frac{1}{250} = 0.004 = 0.4\%$$

51.
$$\frac{3}{250} = 0.012 = 1.2\%$$

52.
$$\frac{333}{500} = 0.666 = 66.6\%$$

53.
$$\frac{2}{9} = 0.\overline{2} = 22.\overline{2}\%$$

54.
$$1\frac{4}{5} = 1.8 = 180\%$$

55.
$$\frac{7}{5000} = 0.0014 = 0.14\%$$

56.
$$(2\frac{1}{5})(5\frac{4}{9}) = \frac{11}{5} \cdot \frac{49}{9} =$$
$$\frac{539}{45} = 11\frac{44}{45}$$

57.
$$3x + \frac{4x}{4} = 3x + x = 4x$$

58.
$$x^4 \cdot x = x^5$$

59.
$$(3xy^2)(-4x^2) = -12x^3y^2$$

245

60. $60xy^2 + 15x^2y$
    $2 \cdot 2 \cdot 3 \cdot 5 \cdot x \cdot y \cdot y +$
    $3 \cdot 5 \cdot x \cdot x \cdot y$
    $15xy(4y + x)$

61. $0.000000837 = 8.37 \times 10^{-7}$

62. $9.378 \times 10^8 = 937,800,000$

63. $(3x^2y)^3 = 27x^6y^3$

64. $\dfrac{4}{3x} + \dfrac{5}{4x} = \dfrac{16}{12x} + \dfrac{15}{12x} = \dfrac{31}{12x}$

65. $9(x - 6) + 7(8 - x) =$
    $9x - 54 + 56 - 7x = 2x + 2$

66. $(x + 2)(x - 5) =$
    $x^2 + 2x - 5x - 10 = x^2 - 3x - 10$

67. $(4a^3 + 2a^2 - 7) + (9a^2 + 4a + 9) =$
    $4a^3 + 2a^2 - 7 + 9a^2 + 4a + 9 =$
    $4a^3 + 11a^2 + 4a + 2$

68. $\dfrac{19x^2y}{38x^3y^2} =$

    $\dfrac{19 \cdot x \cdot x \cdot y}{2 \cdot 19 \cdot x \cdot x \cdot x \cdot y \cdot y} = \dfrac{1}{2y}$

69. $3\sqrt{81} - 4\sqrt{36} = 3(9) - 4(6) =$
    $27 - 24 = 3$

70. $\dfrac{2xy}{9} \cdot \dfrac{18x^3}{y^2} = \dfrac{4x^4}{y}$

71. $(0.25)(4)(\dfrac{2}{9}) = \dfrac{1}{4} \cdot \dfrac{4}{2} \cdot \dfrac{2}{9} = \dfrac{2}{9}$

72. Third

73.
$$3x - 13 < 14$$
$$\quad\;\; 13 \quad 13$$
$$3x < 37$$
$$\dfrac{3x}{3} < \dfrac{27}{3}$$
$$x < 9$$

74. $2x + y = 7 \quad x = 4; y = -1$
    $y = 7 - 2x$
    $y = 7 - 2(4)$
    $y = 7 - 8$
    $y = -1$

   $y = 7 - 2x \quad x = 1; y = 5$
   $y = 7 - 2(1)$
   $y = 7 - 2$
   $y = 5$

   $y = 7 - 2x \quad x = 0; y = 7$
   $y = 7 - 2(0)$
   $y = 7 - 0$
   $y = 7$

75. $60\% \text{ of } 450 = x$
    $(0.6)(450) = x$
    $270 = x$
    There are 270 freshmen in the class.

76. $R \cdot T = D$
    $(12.5)(2\dfrac{1}{2}) = D$
    $(12.5)(2.5) = D$
    $31.25 \text{ miles} = D$

246

77.   6 nickels and 13 dimes

value	# of coins	total
10 cents	$x + 7$	$10(x + 7)$
5 cents	x	$5x$

$$10(x + 7) + 5x = 160$$
$$10x + 70 + 5x = 160$$
$$15x + 70 = 160$$
$$-70 = -70$$
$$15x = 90$$
$$\frac{15x}{15} = \frac{90}{15}$$
$$x = 6$$

78.   16 dimes and 32 quarters

value	# of coins	total
10 cents	x	$10x$
25 cents	$2x$	$50x$

$$10x + 50x = 960$$
$$60x = 960$$
$$\frac{60x}{60} = \frac{960}{60}$$
$$x = 16$$

79. $R \cdot T = D$

$350 \cdot x =$

$400 \cdot x =$ $\Big\rangle$ 4000 ft.

$$350x + 400x = 4000$$
$$750x = 4000$$
$$\frac{750x}{750} = \frac{4000}{750}$$
$$x = 5\frac{1}{3} \text{ minutes}$$

It will take Lori and Peter $5\frac{1}{3}$ minutes.

80.   28% of 245
(0.282)245 ≈ 70 live with parents

63.3% of 245
(0.633)245 ≈ 155 live in dorms

2.0% of 245
(0.02)245 ≈ 5 live in fraternities

6.5% of 245
(0.065)245 ≈ 16 live in private homes

APPENDIX A: THE ARITHMETIC OF WHOLE NUMBERS

A.1 Reading and Writing Numbers

1. Five thousand, 5000

2. Seven thousand, 7000

3. Sixty, 60

4. Thirty, 30

5. Thirteen, 13

6. Seventeen, 17

7. Thirty-five, 35

8. Sixty-two, 62

9. Eight hundred four, 804

10. Six hundred two, 602

11. One million, six hundred fifty-three thousand, nine hundred seventy, 1,653,970

12. Seven million, eight hundred forty-five thousand, seven hundred five, 7,845,705

13. 10, ten

14. 20, twenty

15. 507, five hundred seven

16. 408, four hundred eight

17. 6000, six thousand

18. 2000, two thousand

19. 25,605, twenty-five thousand, six hundred five

20. 32,804, thirty-two thousand, eight hundred four

21. 5,000,000,000, five billion

22. 6,000,000,000, six billion

23. 17,852,027, seventeen million, eight hundred fifty-two thousand, twenty-seven

24. 24,739,054, twenty-four million, seven hundred thirty-nine thousand, fifty-four

A.2 Addition

1.
$$\begin{array}{r} 2 \\ +5 \\ \hline 7 \end{array}$$

2.
$$\begin{array}{r} 5 \\ +4 \\ \hline 9 \end{array}$$

3.
$$\begin{array}{r} 15 \\ +32 \\ \hline 47 \end{array}$$

4.
```
  23
+71
  94
```

5.
```
  7
+9
 16
```

6.
```
  8
+7
 15
```

7.
```
  35
+46
  81
```

8.
```
  47
+24
  71
```

9.
```
  1
  5
+9
 15
```

10.
```
  2
  6
+8
 16
```

11.
```
  3
  7
  6
+5
 21
```

12.
```
  4
  8
  7
+6
 25
```

13.
```
  91
+19
 110
```

14.
```
  19
+91
 110
```

15.
```
  991
+119
 1110
```

16.
```
  911
+199
 1110
```

17.   2134 + 8976 + 11234 + 566 = 22,910

18.   2845 + 7832 + 54113 + 744 = 65,534

19.
```
   1,111
     234
   2,987
  10,555
 +8,000
  22,887
```

20.
```
   2,222
     567
   6,754
  30,555
  +5,000
  45,098
```

A.3 Subtraction

1.
```
    5
   -4
    1
```

2.
```
   19
   -8
   11
```

3.
```
   38
  -14
   24
```

4.
```
  177
  -64
  113
```

5.
```
   32
  -17
   15
```

6.
```
   27
  -19
    8
```

7.
```
  123
 -109
   14
```

8.
```
  546
 -517
   29
```

9.   51,234 − 21,600 = 29,634

10.   72,500 − 13,982 = 58,518

11.   100,000 − 9,199 = 90,801.

12.   1,500,000 − 259,768 = 1,240,232

A.4 Multiplication

1.
```
   21
 x  4
   84
```

250

2.
```
    32
   x 3
    96
```

3.
```
    48
   x 6
   288
```

4.
```
    59
   x 7
   413
```

5.
```
     74
   x 21
     74
   1480
   1554
```

6.
```
     81
   x 31
     81
   2430
   2511
```

7.
```
     93
   x 48
    744
   3720
   4464
```

8.
```
     67
   x 56
    402
   3350
   3752
```

9.
```
    81
   x 0
     0
```

10.
```
    72
   x 0
     0
```

11.
```
    153
   x 32
    306
   4590
   4896
```

12.
```
    261
   x 23
    783
   5220
   6003
```

13.
```
    15
   x 8
   120
```

14.
```
    27
   x 9
   243
```

15.
```
    64
   x 7
   448
```

16.
```
    98
   x 6
   588
```

17. $\begin{array}{r} 100 \\ \times\ \underline{5} \\ 500 \end{array}$

18. $\begin{array}{r} 200 \\ \times\ \underline{7} \\ 1400 \end{array}$

19. $\begin{array}{r} 53 \\ \times\ \underline{20} \\ 1060 \end{array}$

20. $\begin{array}{r} 72 \\ \times\ \underline{10} \\ 720 \end{array}$

21. 5281 x 1234 = 6,516,754

22. 8964 x 2541 = 22,777,524

23. 100,005 x 785 = 78,503,925

24. 346,200 x 958 = 331,659,600

A.5 Division

1. $10 \div 5 = 2$

2. $12 \div 6 = 2$

3. $15 \div 5 = 3$

4. $18 \div 9 = 2$

5. $0 \div 10 = 0$

6. $0 \div 8 = 0$

7. $22 \div 0 =$ undefined

8. $36 \div 0 =$ undefined

9.
$$\begin{array}{r} 6 \\ 13\overline{)78} \\ \underline{78} \end{array}$$

10.
$$\begin{array}{r} 6r2 \\ 15\overline{)92} \\ \underline{90} \\ 2 \end{array}$$

11.
$$\begin{array}{r} 7r6 \\ 21\overline{)153} \\ \underline{147} \\ 6 \end{array}$$

12.
$$\begin{array}{r} 7r13 \\ 23\overline{)174} \\ \underline{161} \\ 13 \end{array}$$

13.
$$\begin{array}{r} 30 \\ 20\overline{)600} \\ \underline{60} \\ 00 \end{array}$$

14.
$$\begin{array}{r} 23r10 \\ 30\overline{)700} \\ \underline{60} \\ 100 \\ \underline{90} \\ 10 \end{array}$$

252

15.
```
        40r5
   15)605
      60
      05
      00
       5
```

16.
```
        55
   15)825
      75
      75
      75
```

17.   1785 ÷ 17 = 105

18.   2821 ÷ 13 = 217

19.   16,422 ÷ 51 = 322

20.   19,920 ÷ 48 = 415

21.   322,091 ÷ 6853 = 47

22.   545,702 ÷ 1942 = 281